The**Youth**Learn Guide

A Creative Approach to Working With Youth and Technology

The YouthLearn Guide and the YouthLearn Initiative were created by the Morino Institute.

Education Development Center, Inc.
55 Chapel Street
Newton, MA 02458-1060
1-800-449-5525
www.edc.org

Design by Particle
Rockville, MD
www.particlefactory.com

Photography:
Center for Educational
Design and Communication
Washington, DC
www.cedc.org

ISBN 0-89292-267-2

PREFACE

Over the past several years, youth programs all over the country have sought new ways of incorporating computers, the Internet, and other new technologies into their work. What they have discovered is that it takes much more than good technology and good intentions to achieve results. For new technologies to live up to their full promise, it also takes a tremendous amount of careful planning, creativity, training, and support.

With this *YouthLearn Guide*, we hope to help you get started on the right foot in your own community. Unlike the frustrating manual that came with your computer, this one is a hands-on, easy-to-understand guide. We hope you will make it golden with heavy highlighting and keep it by your side as you go about building and refining creative and exciting learning programs.

In working to improve the lives of children in low-income communities, we at the Morino Institute have participated in several dozen initiatives to bring the benefits of technology to a wide range of youth-serving nonprofit organizations. The YouthLearn Initiative—including this guide, the comprehensive YouthLearn website (www.youthlearn.org), and the fast-growing YouthLearn online community—grew out of one of these valuable experiences. In 1998, the Morino Institute joined with four respected community organizations in Washington, DC, each with a successful after-school program and a dynamic leader committed to giving children new ways to learn, in launching the Youth Development Collaborative Pilot. This comprehensive two-year project was intended to help these partner organizations establish high-quality technology learning centers as a core component of their youth development programs.

The goal of these technology learning centers was not to teach technology as such but rather to use technology to spark project-based learning—that is, to inspire collaborative learning activities that enable students to connect their classroom work or after-school activities with real-life experiences and concerns. To offer just one recent example, one partner organization worked with a group of young people who used resources they found on the Internet to test the quality of the water flowing into their homes and in the nearby Anacostia River. After writing up their results on computer spreadsheets, the students compared what they found with government water-quality standards. Then, using the Internet and their newly stoked curiosity about the chemistry of water, they began to learn about the Clean Water Act, which was pending reauthorization in Congress. Many of the students lived within two miles of the Capitol, but this was the first time they felt connected to the work inside that ornate domed building—and the first time they realized that they could become active participants in their community.

Just as this guide documents and celebrates the power of collaboration, it also marks the beginning of a new collaborative venture for the YouthLearn Initiative. In December 2001, YouthLearn moved its institutional home from the Morino Institute to Education Development Center, Inc. EDC has a long history of work dedicated to youth development, project-based learning, and expanding access to technology. Together, EDC and the Morino Institute will help YouthLearn bring more ideas, resources, and knowledge to everyone who works with and cares about young people.

Let no one try to convince you that introducing technology into a learning program is a piece of cake. It isn't! But the step-by-step lessons, worksheets, guidelines, and other tools in this manual will show you how it can be done, and done well. If it's done well—and supported with the necessary resources—you can have a profound effect upon the lives of those you reach. You can make learning relevant rather than remote. You can help engage students who have had little success with traditional textbooks in large classrooms. You can give young people in isolated inner cities and remote rural areas the sense that they are connected to a much larger community. It's hard to integrate new technologies into learning programs, but it's worth it. We thank you for your interest in taking up the challenge.

Mario Morino and Tracy Gray
Morino Institute

ACKNOWLEDGMENTS

The YouthLearn Guide represents the work of many dedicated and talented people who have contributed to the Morino Institute's YouthLearn Initiative. A critical component of the initiative was the Youth Development Collaborative (YDC) Pilot. Through this effort, the Institute worked directly with four diverse community-based organizations to help them plan and apply technology to strengthen their organizations and programs:

- Calvary Bilingual Multicultural Learning Center

- Community Preservation and Development Corporation

- Friendship House Association

- Perry School Community Services Center

These organizations were selected because their leaders have consistently taken a strategic, long-term view of how best to address the challenges facing the children and families in their communities. We are grateful to these leaders for taking the time to help us understand their work and for their contributions to the success of the pilot:

- Beatriz (BB) Otero, Executive Director, Calvary Bilingual Multicultural Learning Center

- Leslie Steen, CEO, President, and Treasurer, Community Preservation and Development Corporation

- Donald Hense, President and CEO, Friendship House Association

- Paul McElligott, Executive Director, Perry School Community Services Center

And our heartfelt thanks go to their staff, who demonstrated their deep commitment to creating innovative learning experiences for the children and youth they serve: Steven Berry, Marion Brown, Albert J. Browne, Louis Davis Jr., Jorge Frederick, Kelly Gainer, Tomeka Gibbs, Jomo K. Graham, Marta Jiménez, Daryl Johnson, Nancy Morales, Rafiki Morris, Andrea Neville, Elaine Richardson-Dalzell, Eric Rowe, Rosa Shelton, Marta Urquilla, PJ Urquilla, Troy Wolfe, and Kendra Wymes.

The primary writers of this guide, Janis Cromer and Lorin Driggs, brought rich experience to their task—Janis from her educational reform efforts with the DC public schools and Lorin from her work with the New York Times Learning Network and Bank Street College. Together they did an outstanding job of turning volumes of documents, transcripts, lesson plans, ideas, and concepts created during the YDC Pilot into this user-friendly manual. A special note of thanks goes to our patient and diligent editors, Katarina Rice and Caroline Polk, who were vigilant in their pursuit of order and consistency throughout the manuscript.

Under the wise leadership of Kit Collins, the staff of the Center for Educational Design and Communication (CEDC)—Angela Marney, Barbara Moreau, Beth Ponticello, and Andrew Tischler—provided invaluable contributions to the pilot and the YouthLearn Initiative, including the inspiring photographs that help bring this guide to life. In her capacity as a senior advisor to the Morino Institute, Kit offered steady direction, guidance, and focus during all phases of the planning and implementation of the pilot and this guide.

Additional thanks go to three expert consultants —Robert Price, Eileen Wasow, and Cornelia Brunner, Ph.D.—who provided imaginative and informative training sessions during the pilot on a wide range of topics, from child development to using interactive media with children and youth.

Finally, our sincere gratitude to the Morino Institute associates whose commitment to excellence and willingness to provide their expertise and support has ensured the quality and integrity of the YouthLearn Initiative. In particular, special recognition goes to Tracy Gray, who provided tireless leadership, direction, and vision throughout this initiative. She was joined by the indomitable Richard McDonnell, who contributed his management skills and humor, and an exceptionally talented team of professionals—Kit Collins, Andrea Schorr, and Candy Taaffe. This team established an innovative, hands-on working relationship with the leaders and staff of the partner organizations to create state-of-the-art learning centers and outstanding out-of-school programs for children. Special recognition goes to Lara Suziedelis Bogle and Lowell Weiss, who were relentless in their determination to ensure the quality and integrity of this guide and its usefulness for a broad audience. Thanks go to Cheryl Collins, who was patient and deliberate about capturing knowledge at each stage of the pilot; Neil Oatley, who took that knowledge and turned it into prose; Victoria Vrana, who led the development of the YouthLearn website and online community; Pam McKeta, who added her expertise to expand these online efforts; and Bob Templin, who provided valuable insight and guidance throughout the initiative. It is worth noting that the networked learning centers would not have been set up without the dedication of two remarkable people, Kim Deane and Connell Jones, who gave real meaning to the term "24/7." In addition, many others at the Institute deserve recognition and thanks for their collective efforts: Peter Bostrom, Don Britton, Bill Coquelin, Leslie Crutchfield, Natalie McPherson, Rick Reeder, and Liz Wainger.

And, of course, our thanks go to the young people who let us into their lives to share in their joy at the discovery of learning in technology-enriched environments.

National Review Panel

CONTENTS

RESOURCES

Worksheets

Samples

Figures

INTRODUCTION

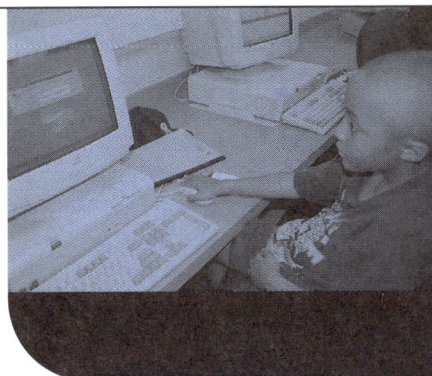

The YouthLearn Guide will help you to:

- Plan a technology-based learning center or program

- Operate and manage the center or program

- Adopt a dynamic, inquiry-based approach to teaching

- Develop project-based learning activities

- Integrate technology into active learning opportunities

- Find additional teaching and learning resources.

Overview

This guide is a resource for planning and implementing creative, active learning centers and programs for children and youth using technology and the Internet. The guide offers practical advice to get your center or program up and running—everything from step-by-step lessons in establishing your vision for your program to tips on sustaining the quality of your program over the long term. The manual also provides teaching and learning materials, including practical, age-appropriate projects and classroom ideas.

The guide grew out of the Morino Institute's experience with community organizations in creating technology-based after-school programs and was designed to provide helpful information regardless of the level of your technical know-how. Whether you are already technologically turbocharged or are just now tiptoeing into the Internet age, the focus of this guide is to help you craft a potent, exciting learning environment for children and youth using technology as a tool toward that goal. The emphasis here is on learning and youth development; the technology is merely a means (albeit a powerful one) to that end.

How to Use This Guide

The guide is not a strict prescription for designing and operating a technology learning program; there is no one right way to do this. Instead, the guide relies on the hard-won experiences of established technology learning centers and presents approaches and techniques that address a wide range of issues common to many technology-based efforts. The strategies have been field-tested and are offered as tips and ideas from which you can choose, depending on your organization's particular needs and interests.

In developing the guide, we at the Morino Institute were aware that no two organizations and no two youth development programs are exactly alike. Similarly, organizations will likely be at different stages in developing their technology initiatives. Therefore, you and your organization may already have tackled the issues presented in some sections of the manual. You are welcome to use those parts of the guide as a refresher or checkpoint. Or, of course, you can skip those sections entirely.

Furthermore, this guide is not a fundraising or technical manual; it won't tell you how to hardwire your computer lab, nor will it help you purchase computers or software (although it does include a brief primer on working with technology and a section on program finances and sustainability). The focus of the guide is to help you use your existing resources, both human and technological, to develop exciting learning opportunities for the young people you serve.

The guide was created for adult practitioners who are working with children and youth in settings that provide learning programs supported by technology tools and the Internet. This group includes leaders of community organizations, program directors, principals, and other educators seeking assistance in creating or managing technology learning programs, as well as teachers, youth development instructors, and classroom aides looking for creative ways to integrate technology into learning activities.

Some information may be more useful for program administrators; other sections may be most helpful for classroom staff. Program directors with responsibilities for both administering technology learning programs and developing instructional content are likely to find relevant material in all sections of the guide. Readers are encouraged to scan the entire guide and focus on areas that are most relevant to their own work. The first page of each chapter of the guide has an "audience box" indicating who might find that particular topic of most value.

The guide was intended to be a hands-on, flexible tool for all users. As a result, each section has a combination of worksheets, checklists, figures, "reality checks," tips, and additional resources for you to apply as you see fit. You are welcome to photocopy, adapt, or share the materials in ways that best suit your program or organization, as long as it is strictly for non-commercial, educational purposes.

> " *The emphasis here is on learning and youth development; the technology is merely a means (albeit a powerful one) to that end.* "

Overview

According to the Pew Partnership for Civic Change, the parents of more than 28 million children work outside the home. Many of those parents are struggling to find high-quality, safe experiences for their children during the after-school hours. After-school programs have a broad range of purposes and missions. However, all types of programs are in great demand. Current research shows that for every two children who need a program, only one slot exists. Even though many parents are willing to pay more for after-school programs, the supply continues to fall short of the demand.[1]

In the past decade, after-school and other out-of-school programs have undergone a transformation and renaissance. What once were primarily drop-in centers that offered a safe place for homework help, recreation, and tutoring not only have increased in number but also have changed focus. New federal and local funding for after-school programs began to require an increased educational emphasis, and parents wanted more than supervision for their children—they wanted engaging environments that integrated academic skill-building with activities that promoted the social, emotional, and physical development of their children.

Out-of-School Learning

The explosion in out-of-school offerings has given rise to considerable research about the benefits of such programs. The National Institute on Out-of-School Time engaged in a review of the current research and found that "children who attend high-quality after-school programs have better peer relations, emotional adjustment, conflict-resolution skills, grades, and conduct in school than their peers who are not in after-school programs."[2]

Out-of-school programs differ in their goals and objectives. They also take a variety of forms, including:

- Daycare programs

- After-school programs sponsored by an array of organizations

- School-based, academic, extended-day programs

- Academic enrichment programs not sponsored by schools

- Summer programs and camps.

Contents

- Overview of the state of out-of-school learning

- Core principles of after-school programs

- Assumptions about technology and learning

- Keys to success

Audience

- Youth development staff

- Teachers and center instructors

- Program directors

- Organizational leaders

- Educational administrators

For many programs—particularly after-school programs—a primary goal is to occupy children during the critical hours between 3:00 p.m. and 6:00 p.m. The Pew Partnership has identified three functions for after-school care:[3]

- Supervision

- Enriching programs and experiences and positive social interaction

- Academic improvement.

Program sponsorship also varies. Schools are frequent sponsors, as are nonprofit, for-profit, and religious organizations. Schools have the advantage of continuity, accessibility, resources, and expertise. However, school-run programs also have some disadvantages, such as higher personnel costs (if after-school staff salaries must be equal to teachers' salaries), the possibility of program budget cuts, and children's perception that after-care is an extension of the school day. Community-sponsored programs are generally freer than schools to use innovative curricula and activities to promote student learning. However, staff members may not be adequately trained to provide academic enrichment.

Effective programs, regardless of their structure or sponsorship, share a number of elements:[4]

- Clear goals and outcomes

- On-site management and coordination

- Qualified staff

- A strong focus on safety, health, and nutritional needs of children during program hours

- Effective collaboration with and links to community agencies

- Strong involvement of parents

- Coordination with school-day learning and personnel

- Ongoing evaluation of the program.

Technology and Learning

Over the past decade, technology has become an important learning tool in schools and other traditional educational settings. However, the infusion of technology into out-of-school programs has been somewhat limited.

Although getting wired and equipped with technology is still a formidable challenge in many settings, determining the best *uses* of technology to enhance learning has become a major focus of educational research. Numerous studies have concluded that technology is not a cure-all for education but is a powerful and exciting complement to the key ingredients for learning:

- Skilled, caring teachers and instructors who serve as *coaches* to learning, not as the undisputed dispensers of all knowledge

- An emphasis on *active* learning opportunities that encourage children to *discover* answers and solutions, rather than an emphasis on rote memorization of facts

- Instruction that focuses on *cooperative* or *collaborative* learning among children and youth, rather than on individual, isolated knowledge acquisition.

More Than Boxes and Wires

Throughout this guide, you will encounter the term "technology learning center." We are not using the term as a generic label for any room or set of rooms where children and adults can use computers and Internet connections. Instead, we have a very specific definition in mind, one that makes clear that effective learning requires much more than boxes and wires.

Although the structure of each technology learning center will vary and each center will pursue somewhat different goals, the most successful centers generally share the following attributes:

- The center functions as an integrated component of a comprehensive youth program that provides activities or services in addition to the center's classes (e.g., mentoring, family outreach, and family support services).

- The center's sponsoring organization is a respected and active part of the community, easily accessible to the youth and families served by the center.

- A technology learning center has at least one full-time staff person dedicated to managing and directing the center's classes and associated activities.

- Technology learning center instructors and staff members work closely with other members of the organization's staff to plan and coordinate the entire youth program.

- Both the staff of the technology learning center and the staff of the organization as a whole participate in professional development workshops and other training activities.

- Each technology learning center has a teaching and training space with at least 15 workstations, a high-speed Internet connection, a suite of software appropriate for project-based learning (discussed in detail in Chapters 6 and 7), and peripheral equipment such as digital cameras and printers.

Worksheet 1.1 lists elements that have proven critical to the success of technology learning centers.

"In the past decade, after-school and other out-of-school programs have undergone a transformation and renaissance."

@ For more information, examples, and resources, see the extensive YouthLearn website at www.youthlearn.org

Worksheet 1.1
Keys to Success

This worksheet lists elements that have proven critical to the success of technology learning centers. Use this worksheet to review your plans before you open your doors or to assess your program at various stages of its development and operation. You may want to complete the sections one at a time as you work through the corresponding chapters.

Success Factor	Excellent 1	Very Good 2	Good 3	Fair 4	Needs Improvement 5	Notes/Follow-up
Plan your program (see Chapter 2)						
A group of people dedicated to the success of the program						
A thorough understanding of the organization's capacity and needs related to launching and operating a technology learning center						
A thorough understanding of the needs and desires of the community, including parents, children, staff members, and other stakeholders						
A clear statement of program goals that complements the organization's mission						
Measurable benchmarks for assessing progress toward each goal						
A specific action plan detailing timelines and staff assignments that is updated regularly and serves as a playbook for all staff members						
A flexible program schedule that addresses the needs of all participants and results in a creative, dynamic learning environment						
A plan for regularly reviewing, revising, and improving the program						
A positive, caring learning environment with established norms for behavior and high expectations						

(continued on next page)

Keys to Success

Success Factor	Excellent 1	Very Good 2	Good 3	Fair 4	Needs Improvement 5	Notes/Follow-up
Manage your program (see Chapter 3)						
An organization that is respected and closely tied to its community						
Organizational leadership that is closely involved with the plans and needs of the technology learning center						
Clearly written operational policies and procedures						
Responsible, caring, and skilled staff members						
A plan for continuous professional development and skill enhancement for all staff and volunteers						
Strong fiscal controls						
A support network of partner relationships with other organizations						
An ongoing plan for funding the program (sustainability)						
An established system of regularly assessing progress toward goals, taking corrective action when needed, and fostering the program's continuous improvement						
Staff your program (see Chapter 4)						
The ability to hire and retain a dedicated, skilled core of staff members and volunteers						
Train your staff (see Chapter 5)						
A plan for assessing the skills of the staff and providing regular training to enhance their capabilities						

(continued on next page)

(**Worksheet 1.1** continued)

Keys to Success

Success Factor	Excellent 1	Very Good 2	Good 3	Fair 4	Needs Improvement 5	Notes/Follow-up
Develop project-based learning strategies (see Chapter 6)						
A project-oriented approach to learning						
An inquiry-based approach to projects and activities						
An emphasis on collaboration and cooperative learning among children and youth						
Staff members who understand and use creative lesson plans to structure learning						
Develop effective teaching techniques (see Chapter 7)						
Staff members who instill values and attitudes through use of good modeling techniques						
Use of a variety of teaching techniques to help kids build reading, writing, and thinking skills						
Use of technology as a tool for learning, seamlessly integrated into instruction and activities						
Work with technology (see Chapter 8)						
A staff with a basic understanding of the technical aspects of computers, software, and the Internet						
A solid technology plan to serve as a road map for your organization's technical needs						
A plan for continuously upgrading the technical skills of staff						
A trusted, skilled resource for technical assistance						
An established maintenance plan for the technology learning center's equipment						
A clear policy on Internet safety and security for all participants						

The YouthLearn Guide • Created by the Morino Institute • *www.youthlearn.org*

chapter 2

PLAN YOUR PROGRAM

Overview

Careful planning is critical to the success of your technology learning center. Although not all contingencies can be anticipated through the planning process, the goal of planning should be to create a realistic, workable road map to the creation and operation of a center that meets (and, one hopes, exceeds!) a set of established goals. By having a clearly defined path, you will be much better equipped to assess how unforeseen developments are affecting your plan, make adjustments if necessary, and keep the center's progress on track.

This chapter presents a sequence of planning steps, starting with the formation of a planning team and ending with the development of a schedule for the technology learning center. The chapter also includes a special note on how your planning team can contribute right from the start to the all-important process of building a positive climate for learning.

Contents

● Steps and worksheets for planning a technology learning center from the start

● Guides for assessing need and the organization's capacity

● Forms for program planning and scheduling

Audience

● Organizational leaders

● Program directors

● Educational administrators

✔ Reality Check

When evaluating whether your program has the structure needed to carry out projects, you might ask the following questions, among others:

☐ Does the program have an enrollment process that requires consistent attendance?

☐ Does the program provide transportation, school pick-ups, or other services to help youth get there?

☐ Is the program set up so that youth participating in projects can start and end activities together on regular days or at regular times each week?

☐ Does the program schedule allow for uninterrupted blocks of at least an hour and a half for work on a project?

☐ If additional program time is needed for project work, have parents been asked to allow children to stay longer?

Step 1:
Create a Planning Team

Although you may have done some individual thinking and planning for the center, engaging additional interested people early in the planning process will invariably result in a better-conceived and more widely supported approach as the program develops. A team approach also taps a variety of staff talents and spreads the weight of responsibility around so that no one feels overburdened. An important lesson learned in the Morino Institute's pilot effort to establish four technology learning centers was that the scope of work involved in launching and running an effective center is much broader than for an out-of-school program that does not use technology.

A good planning team is made up of people with a range of skills, expertise, and relationships to the community organization. You may wish to include a parent, volunteer, private-sector representative, or board member on the team. Some planning teams have included a youth representative. Diverse, however, doesn't necessarily mean all-inclusive. A team that's too big won't be as effective as one that's light on its feet—that is, able to tackle problems and act decisively within a reasonable amount of time. Generally, a good size for a planning team is five to seven members. The team also should have a chairperson, usually a staff member in a leadership role within the community organization.

Worksheet 2.1 can help you develop the membership of your technology learning center's planning team.

> "*A good planning team is made up of people with a range of skills, expertise, and relationships to the community organization.*"

Worksheet 2.1
Create Your Planning Team

For each role, list three possible candidates. Your final selection will depend on the person's availability and willingness to serve. Make sure that you have a variety of community roles represented on your team.

Team Role	Where to Look	Candidates
Chairperson	Director of youth development, director of technology learning programs, or other member of the technology learning center's sponsoring organization	
Member	Executive director or other leader of the sponsoring organization	
Member	Teacher or instructor in the community organization	
Member	Financial or development officer from within the organization	
Member	A parent representative	
Member	A representative from a private-sector partner (e.g., a program manager from a business or corporation allied with the community organization)	
Member	A youth representative (e.g., a teen who has participated in the organization's other programs)	

Step 2:
Establish Your Vision and Mission

Rather than plunging into program details, the planning team should use its initial meeting to take a broad view of what it aims to accomplish—sharing ideas about the purpose of the technology learning center, the needs the center will address, the desired outcomes, and so forth. The objective of this initial work is to reach a shared ideal about the technology learning center.

After a period of discussion and debate, this ideal should be crafted into two concise statements. The first is a "vision statement," which should capture dreams and aspirations for the center. The second is a "mission statement," which should articulate your plan of action for achieving your vision. These statements will serve as the compass and the road map for your technology initiative.

Vision Statement

A vision statement is an attempt to articulate the desired future for an organization, center, or program. In the words of management expert Peter Drucker, it is "an organizational dream—it stretches the imagination and motivates people to rethink what is possible."

Worksheet 2.2 provides a list of discussion topics that can assist your planning team in formulating your vision statement. The planning team chairperson can use this worksheet for leading the meeting.

After working through the series of questions in Worksheet 2.2, the chairperson should then move the team to the refinement stage. The emphasis in this stage is on decision making. Refinement involves taking all of the ideas, winnowing out the less workable ones, and ultimately selecting the ones the team intends to pursue. In this case, the goal is for the team to consider all the suggested "pieces" of a vision for the technology learning center and to create a one-sentence vision statement that captures the essence of the group's thinking.

An example might be: "Our vision is that every child served by our center will have access to a world of practical and innovative learning opportunities that will prepare them to reach their full human potential."

Once your center is operational, the vision statement should be posted prominently to remind participants and visitors about the "big picture" purpose of the technology learning center. Posting the statement also helps to inspire program leaders and staff members to stay focused on the importance of that vision as they set their overall priorities and go about their challenging work.

Mission Statement

A well-conceived mission statement provides an orientation for your center. It offers a direction, not a destination. It should define in concrete terms why the center exists. It should be easy to understand. And, in the end, it must give every member of the center's staff a clear sense of what they are there to achieve. An example is: "Our mission is to provide our community with a state-of-the art technology learning center that strengthens the academic skills of children and professional skills of our staff. The center will serve as a vital information hub for people of all ages, to help them meet the critical need for lifelong learning, financial and social services, health and child care, and good jobs."

When your planning team sits down to brainstorm about the components of your mission statement, they should consider the following issues:

- **Current assets.** What factors will contribute to the success of the technology learning center? Examples of current assets include an existing source of funds for the center, a skilled instructor who is already on the staff, or a corps of determined parent volunteers.

- **Current challenges.** What obstacles will the center face in realizing its vision? Such challenges could include a lack of technical assistance in maintaining the learning center equipment, a lack of up-to-date software, or space limitations for the center.

- **Outcomes.** What exactly are you trying to accomplish, and how are you going to accomplish it? How will you measure progress toward your goals? Measuring outcomes on an ongoing basis provides an organization with a steady stream of information and feedback so that it can continually assess its progress in meeting its goals and improving its programs.

Additionally, being able to showcase a program's outcomes attracts new participants, collaborators, and funders.

Worksheet 2.3 will help your planning team address these issues and craft your mission statement.

⚡ Tip

Brainstorming

To elicit the full range of ideas from the planning team during the process of developing vision and mission statements, the chairperson may wish to use a brainstorming technique.

In brainstorming, the goal is to get everyone's creative juices flowing and to come up with as many related ideas as possible, saving judgment for the refinement stage. During a brainstorming exercise, everyone's ideas get written down, and no one is allowed to comment on them, either positively or negatively. The concept behind the technique is to create a fertile environment in which each team member builds on the thoughts of others without being sidetracked by prematurely evaluating the practicality of the ideas.

For example, the chairperson might open the discussion by posing the question "In two years, what

will the technology learning center look like and be doing?" The question can be posted on a large sheet of newsprint, and the chair or another team member can serve as the recorder to jot down key phrases or comments made by each team member. The goal is to write down as many ideas and options as possible without pausing to critique any particular idea or suggestion.

Using graphic organizers is another technique that may be helpful during the planning process. Also referred to as "visual maps" and "idea maps," among other names, graphic organizers use simple visual templates to help people generate, collect, organize, and record ideas. Some templates are best for the brainstorming phase, others work best for refinement, and some are equally useful in both stages. Examples of graphic organizers can be found in Chapter 7.

Worksheet 2.2
Establish Your Vision

Use the questions below to guide a discussion among key planning team members to identify the dreams and aspirations for your center. After a period of brainstorming, you should move to develop a concise statement that will present your guiding vision to the entire community.

Create Your Vision Statement

Discussion Topics	Ideas From Participants
What are the major needs of our community that we would like to address?	
How will we use technology to improve the quality of life in our community?	
What will our technology learning center look like?	
Who will be served by our technology learning center?	
Ten years from now, what will be different in our community as a result of our efforts?	
What is our vision statement? (Your vision statement should be one or two sentences. It should motivate and inspire others to become involved with your center.)	

The YouthLearn Guide • Created by the Morino Institute • *www.youthlearn.org*

Worksheet 2.3
Establish Your Mission

Once you have developed a vision statement (see Worksheet 2.2), you will be ready to create a mission statement. Your mission statement should provide a brief, compelling, and realistic plan for fulfilling your vision for the center.

Create Your Mission Statement

Discussion Topics	Ideas From Participants
What current assets (physical space, hardware, software, staff, funds, etc.) are available for our center?	
What type of learning programs and services will we offer to children and their families?	
What obstacles will the center have to overcome in order to be successful?	
What are the intended outcomes for children and their families, and how will we measure these outcomes?	
What is our mission statement? (Your mission statement should be two or three sentences. It should summarize the purpose and intended outcomes for your center.)	

Step 3:
Assess Your Organization's Capacity

After the planning team has spent some time considering the "ideal" technology learning center and creating a vision for the program, it will be important to inject a dose of reality before plunging too far ahead. Having unrealistic expectations that then go unfulfilled almost guarantees a program's failure.

One way to keep a realistic perspective is to conduct a capacity assessment—a review of everything the technology learning center will need in order to function efficiently and the capacity of your organization to provide the resources to meet those needs. Conducting the capacity assessment can be a job that the entire planning team tackles, or it can be assigned to a subset (two or three members) of the team.

Network With Others

A first step in conducting a capacity assessment may be to take a close look at other technology-based programs to see the range of resources and factors that had to be in hand or acquired over time. Ideally, members of the planning team should visit some programs and arrange to spend time with the program leaders to hear issues and solutions directly from other frontline technology learning center pioneers. If onsite visits are not possible, telephone conversations with leaders of other programs and participation in online learning groups would be a sound investment of time.

By knowing what's out there and who's doing what, you can:

- Build on existing work rather than waste time and money duplicating efforts

- Get practical help from those who have been there before

- Identify sources of information, training, and funding

- Build a support system and, perhaps, a steady supply of continuing technical assistance and advice for your program as it matures.

Evaluate Your Capacity

After the planning team members gain some familiarity with the range of capabilities and resources needed to develop a successful technology learning center, they should examine the organization's capacity to address those needs.

The following sections describe some capacity issues that the planning team should explore.[5] You can then use Worksheet 2.4 to develop your own capacity assessment and identify possible strategies for overcoming any current limitations your organization might have.

- **Environment and facilities.** Are the indoor and/or outdoor spaces for the technology learning center sufficient for the kinds of activities you would like to do? Are there any constraints to those spaces that would preclude certain activities? For instance, does the size of the rooms available for the center limit the number or ages of children your program can serve?

  Are there places for keeping projects that are still in the works? Is there space to store materials and reference works? If children can't leave a project and get back to it in the next session, that will limit the kinds of activities your center can undertake.

- **Equipment and materials.** What equipment and materials are available? What is their condition? What other equipment or supplies will be needed? Is there a resource (internal or external) to troubleshoot equipment problems and do routine maintenance? Is there enough money available to buy the materials your activities will require?

- **Child dynamics.** How many children do you expect to have in your program, and what is the age range and the gender mix? Do you anticipate any behavioral problems that could affect the success of your chosen activities? If so, think about ways to address such problems (e.g., dividing the group into several smaller groups or planning activities that allow for a wider range of abilities and ages). Will the children have language or learning needs that require special training or skills on the part of instructors?

- **Administration, staff, and community.**
How committed are the staff and the community to the program? Do staff members feel they have a stake in its success? Do you have top-level leadership from the sponsoring organization? If the support of the administration or staff is lacking, you will need to address the reasons why they are not supportive. Do you have enough personnel to staff the program successfully? Are volunteers available to assist?

- **Instructional skills and content knowledge.**
Do the activities you want to do require specific knowledge on the part of instructors? Do you have a plan and the financial resources to provide training to increase staff members' instructional skills and content knowledge?

- **Time.** What time considerations must you factor into the structuring of your program? How will those issues affect the kinds of projects your technology learning center can do?

"Having unrealistic expectations that then go unfulfilled almost guarantees a program's failure."

Worksheet 2.4
Assess Your Capacity

Assessing your program's "capacity"—or capabilities—is the foundation of the planning process. In each of the categories below, rate your current level of capacity and think about ways to improve in areas where you have identified limitations.

Technology Learning Center Component	What is our capacity in this area?			Notes/Possible Solutions to Limitations
	Excellent	Adequate	Some Limitations	
Environment and Facilities				
Room for the anticipated number of children				
Regular access to the computers and other equipment				
Friendly, open spaces for projects				
Storage space				
Equipment and Materials				
Quantity and condition of equipment				
Quality and quantity of software				
Quantity of supplies				
Resources to acquire needed supplies				
Provisions for equipment troubleshooting and maintenance				
Child Dynamics				
Plan for addressing problems related to the number, age range, gender mix, or behavior of children in the program				

(continued on next page)

The YouthLearn Guide • Created by the Morino Institute • www.youthlearn.org

Assess Your Capacity

Technology Learning Center Component	What is our capacity in this area?			Notes/Possible Solutions to Limitations
	Excellent	Adequate	Some Limitations	
Administration, Staff, and Community				
Strong administrative leadership support of the program				
Strong program support from parents and other community members				
Enough skilled staff members to run the program				
Plan for recruiting and retaining additional staff members, if needed				
Availability of volunteers and support personnel				
Instructional Skills and Content Knowledge				
Talented and creative instructors				
Instructors with appropriate content knowledge				
Plans and resources to provide ongoing staff training				
Time				
Hours of operation appropriate for program needs				
Flexibility in scheduling				

Adapted from the wNet School workshop After School Programs: From Vision to Reality
(www.thirteen.org/wnetschool/concept2class/month11/implementation.html).

Plan Your Program

Step 4:
Assess Your Constituents' Needs

Before you delve into planning the details of your program, it is important to gather opinions and input about the proposed technology learning center from a variety of "stakeholders"—the people who have a direct connection to the program. For out-of-school programs, stakeholders include staff members, parents, children, and other community members. What would students like to learn or do in a technology-based program? What are the specific interests of your staff? What would parents like to see offered in such a program? It is also important to consider the needs, wishes, and distinct culture of the community the center will serve. For example, are the children or youth in your program from families who are recent immigrants to this country? Do community members want or need particular areas of instruction?

"Before you delve into planning the details of your program, it is important to gather opinions and input from a variety of stakeholders."

Why Conduct an Assessment?

Gauging stakeholders' desires and needs related to a technology learning center—also known as conducting a needs assessment—serves several important purposes:

- It helps ensure that the goals and planned activities for the technology learning center are aligned with the desires and expectations of all the people who will be closest to the program.

- It provides information about the level of knowledge, skills, and interests among the different stakeholders, which will be valuable in determining resource allocations and targeting activities to identified needs. (For example, assessing the staff members' experience with technology will help decide how training resources should be directed.)

- It provides a means of getting the word out about the creation of the center.

- It builds wider support for the technology learning center because its development will be based in part on the input from actual stakeholders.

How to Assess

You can conduct a needs assessment in several different ways. You may wish to employ different methods for various groups, depending on which audience you are soliciting input and opinions from. Information collection strategies include:

- Individual interviews with a representative sample of the various stakeholder groups

- Telephone interviews or surveys

- Focus groups (stakeholders brought together in small groups that meet for 60 to 90 minutes each to discuss predetermined questions)

- Written surveys.

One approach might be to conduct a few individual interviews with key community leaders; hold a focus group session for staff members, supplemented by a written survey to be completed and submitted after the session; have a round of telephone interviews with a small sample of parents; and, finally, hold a group discussion with children.

Regardless of the methods employed, it is important to do the following:

- Prepare and write down all questions in advance for interviews, focus groups, and telephone surveys.

- Use the same questions for all members of a stakeholder group. For example, the questions you ask one instructor should be used again when you interview other instructors.

- Make sure each question is focused and not too broad; otherwise, you will end up with confusing information. For example, if you ask, "Would you like to see the program offer children opportunities to learn keyboard skills and creative uses of a digital camera?" respondents might answer yes—but is their primary interest in the keyboard skills or use of the digital camera? Instead ask, "Which do you think is more important for the program to offer, instruction in keyboarding or instruction in using a digital camera creatively?"

- Use language that each of your stakeholder groups can easily understand. Avoid jargon, overly technical words, and acronyms.

- Have a means of recording the information for later reference. For instance, have someone take notes during interviews or group discussions.

See Worksheet 2.5 for one approach to gathering information systematically.

💡 Tip

A Special Note Regarding Children

You have a lot to gain from listening to all of your program's stakeholders—especially the children who will participate in your program. Even very young children, through guided question-and-answer sessions, can provide information that will be valuable to your planning activities. The level of participation in your program's activities and the effectiveness of those activities in achieving program goals will be greatly enhanced if what you offer reflects both the needs and the interests of the children.

When you're soliciting input from children and youth, make it clear to them that you are keenly interested in everyone's ideas and opinions and will take them all seriously; however, also make it clear that you may not be able to put all those ideas into action. It is up to the adult leaders in the program to establish the ground rules, guidelines, and priorities that determine how the children's input is used.

Always document the ideas and suggestions of children so that you can review them thoughtfully with the planning team or other program staff. Sometimes a little creative problem solving can turn an initially improbable idea into a valuable feature of your program. That's why it's important not to dismiss children's suggestions out of hand.

Listening to your participants has at least two additional benefits. First, it boosts their sense of ownership in the program. Second, it models a collaborative, respectful, and inclusive approach to working together that they may be able to draw on in other situations.

Worksheet 2.5
Create a Parent Survey

Use this sample survey as a starting point to design a survey for parents and other constituents of your technology learning center. Add or delete questions according to your need for information.

Questions for Parents and Other Community Members

[Insert an introductory or background statement, such as "The XYZ organization has an opportunity to create a new after-school offering that aims to use various types of technology, such as computers and digital cameras, in a creative learning program. The program would be open to children ages 5 to 13 from 3:00 to 6:00 p.m. on weekdays. This program is in the early planning stages, and we would like to get your ideas and suggestions."]

1. Do you think your child/children would participate in a technology learning program? (Check one.) ☐ Yes ☐ No

2. Do you have a computer at home? ☐ Yes ☐ No

3. If so, does your child/children use the computer? ☐ Yes ☐ No

4. In what ways does your child/children use the computer? (Check all that apply.)

   ☐ Games
   ☐ Email with friends/family
   ☐ Making web pages

   ☐ Using the Internet for homework research
   ☐ Online chat with friends
   ☐ Listening to music

   ☐ Other _____
   _____

5. What types of activities do you think our program should offer? (Technology learning can be incorporated into many kinds of activities.) Rank your top eight choices from 1 to 8, with 1 as your top choice. [Note: Be sure to list activities that reflect what your center will realistically be able to offer; leave lots of room for "Other."]

   ___ Chess
   ___ Community service
   ___ Homework help or tutoring
   ___ Photography
   ___ Poetry writing
   ___ Newspaper
   ___ Saturday academy
   ___ Science experiments
   ___ Aerobic exercise
   ___ Basketball

   ___ Martial arts
   ___ Soccer
   ___ Volleyball
   ___ Arts and crafts
   ___ Card and board games
   ___ Drama
   ___ Field trips
   ___ Music
   ___ Peer counseling/conflict resolution

   ☐ Other _____
   _____

   ☐ Other _____
   _____

   ☐ Other _____
   _____

6. What is the most you would be willing or able to pay per child for a technology learning activity that meets five times a week for 16 weeks? (Check one.)

   ☐ I am not willing/able to pay for such activities.
   ☐ Less than $75
   ☐ $76 to $150
   ☐ $151 to $225
   ☐ More than $225

(continued on next page)

Worksheet 2.5 continued

Create a Parent Survey

7. Please indicate the days and times that you would like your child/children to be able to attend out-of-school activities. (Check all that apply.)

☐ Monday ☐ Tuesday ☐ Wednesday ☐ Thursday ☐ During school vacations

☐ Friday ☐ Saturday ☐ Sunday ☐ During the summer

8. Would your child/children need a ride home after an activity?

☐ No, I would be able to pick up my child/children after an activity.

☐ Yes, I would need the program to provide transportation for my child/children after an activity.

9. Are you interested in volunteering to help with the program? (Check one.) ☐ Yes ☐ No (go to question 11)

10. In what ways would you like to volunteer? (Check all that apply.)

☐ Teach a class

☐ Help a teacher with a class

☐ Help with paperwork (e.g., keep attendance, fill out forms)

☐ Publicize the program (e.g., write for the newsletter, pass out flyers)

☐ Greet participants and answer questions

☐ Provide help wherever needed

☐ Other (please specify)_____

11. Personal information (optional):

_____   _____
(Your name)   (Address)

_____   _____
(Home telephone)

Best time to reach you at this telephone number _____

Would you like to receive more information about the program? ☐ Yes ☐ No

Your child's/children's name(s) and grade(s):   _____ Grade _____
                                                (Name)

                                                _____ Grade _____
                                                (Name)

                                                _____ Grade _____
                                                (Name)

Thank you!

Adapted from K. E. Walter et al., Beyond the Bell: A Toolkit for Creating Effective After-School Programs, Second Edition
(Naperville, IL: North Central Regional Educational Laboratory, 2001), p. 71 (www.ncrel.org/after/bellkit.pdf).

Step 5:
Establish Goals and Benchmarks

Establishing clearly articulated goals and benchmarks (i.e., interim steps toward meeting the goals) and revisiting them periodically is a particularly critical step in the planning process. After reviewing and digesting the information gathered through the capacity and needs assessments, the planning team can now use that knowledge, along with the established vision for the program, to develop specific goals for all aspects of the technology learning center.

Clearly defined goals serve a number of purposes. They help keep staff members and leaders focused on the aims of the program. At the same time, goals help parents, youth, funders, and other program stakeholders understand why certain decisions are being made in the design and implementation of the program. Goals also serve as the basis for measuring the progress of the technology learning center and can be a reference point for making needed changes.

Goals may be broad statements, but they should be measurable and directly related to program activities. "Become the best technology learning center in the state" may be an admirable objective, but it is a difficult goal to assess (how would you determine whether it is the "best"?), and it does not provide a direction or course of action for program activities. Goals also should have an established target date.

Benchmarks are more specific; they are the established milestones along the path to each goal. They, too, should be measurable and provide interim checkpoints for evaluating progress toward the goals. Like goals, benchmarks should have specific time frames.

To assist you in developing goals and benchmarks for your center, the following pages contain two worksheets. Worksheet 2.6 can be used with the planning team to establish goals and target dates for the various components of your program. Once you've identified your program goals, you can use Worksheet 2.7 to develop benchmarks for each goal.

Examples of Goals	Related Benchmarks
Develop and launch a comprehensive outreach campaign for identifying and recruiting students for the program.	• Establish collaborative relationship with schools and major community- and faith-based organizations prior to launch. • Identify and train 10 recruitment volunteers from the community.
Enroll 200 children by the end of the first year of the program.	• Pre-register 30 students for the program. • Increase attendance numbers by 5 percent each month.
Launch website with online registration and student tracking capacity.	• Beta-test website with administrators, staff, and 25 potential users from the community. • Train all staff in the use of a tracking system.

Worksheet 2.6
Set Your Goals

This worksheet can be used to record both intermediate and long-term goals. You may want to use it with your entire planning committee to solicit a longer list of goals and then, through discussion and refinement, narrow the list. In the first column, write general goal areas, such as staff development, program operations, participant outcomes, community support, fundraising, or any other categories for which you want or need to establish goals. In the second column, list up to four goals for that program component. Finally, in the third column, set a target date for accomplishing each goal.

Program Area	Specific Goals	Target Date
Program Operations and Project Design	a.	
	b.	
	c.	
	d.	
Staffing and Training	a.	
	b.	
	c.	
	d.	
Participant Outcomes	a.	
	b.	
	c.	
	d.	
Community Awareness and Support	a.	
	b.	
	c.	
	d.	
Fundraising	a.	
	b.	
	c.	
	d.	
Other		

Adapted from K. E. Walter et al., Beyond the Bell: A Toolkit for Creating Effective After-School Programs, Second Edition (Naperville, IL: North Central Regional Educational Laboratory, 2001), pp. 13-14 (www.ncrel.org/after/bellkit.pdf).

Worksheet 2.7
Establish Benchmarks

Use this worksheet as a template to help you determine benchmarks for your goals. Write one of your goals in the first column, and list the possible benchmarks for that goal in the next column. Then, for each benchmark, answer the two questions. Finally, in the last column, list the actions that need to be taken in order to reach the benchmark. Use a new copy of the worksheet for each of your goals.

Program Goal	Possible Benchmarks	Questions to Ask About Each Benchmark		Actions to Be Taken
		Is the bench-mark directly related to the goal?	How will the benchmark be measured? List data sources—e.g., attendance rate, amount of new funding, staff surveys. (Note: If you cannot measure the bench-mark, you should replace it.)	
	1.			
	2.			
	3.			

Adapted from K. E. Walter et al., Beyond the Bell: A Toolkit for Creating Effective After-School Programs, *Second Edition (Naperville, IL: North Central Regional Educational Laboratory, 2001), pp. 125-26 (www.ncrel.org/after/bellkit.pdf).*

Step 6:
Develop an Action Plan

With the creation of goals and measurable benchmarks, perhaps the most detailed aspect of planning is at hand—the development of an action plan. This is the "playbook" or road map for implementing your program, and it is a vital management tool. The action plan

- Identifies and prioritizes the things that need to be done

- Assigns specific tasks to specific people

- Sets due dates for completion of tasks

- Provides a comprehensive view of all activities to avoid setting unreasonably tight or impossible deadlines.

An action plan is a "living" document, a unified reference point for everyone involved in creating and running the program. The best action plans extend no more than 90 days into the future. They should be reviewed and updated weekly or every other week to track progress and highlight emerging problems.

For each major aspect of your program, the action plan should list the specific tasks to be completed, who has the lead responsibility for each task, which staff members will be assisting with the task, when the task will be completed, and any notes regarding the task. Examples of tasks are "Finalize staff hiring" and "Complete the installation of new software."

Worksheet 2.8 is a template for creating an action plan for your technology learning center.

> "*An action plan is the 'playbook' or road map for implementing your program.*"

Worksheet 2.8
Create an Action Plan

Use this worksheet as a template for creating an action plan. Your actual plan will be several pages long, probably devoting a page or two to each major program component.

Action	Person With Lead Responsibility	Other Assigned Staff	Target Date	Comments and Notes
EXAMPLE				
Staffing				
Task 1: Hire a technology director				
1.1 Develop the job description				
1.2 Advertise the position in newspapers, on the Internet, etc.				
1.3 Review applications and recommend top candidates				
1.4 Interview candidates				
1.5 Check references				
1.6 Make the job offer				

Step 7:
Create a Program Description

Many different kinds of people—including potential funders, parents, job seekers, potential volunteers, and representatives of the media—will want to understand what your program is all about. By creating a concise (usually one-page) program description, you will have a ready source of information about your technology learning center for all audiences. A program description contains the "official" facts about your technology learning center. It should answer fundamental questions about when and why the program was created, what it consists of, what its goals are, and what accomplishments it may have achieved. A good program description serves not only as a valuable tool for communicating with interested outsiders but also as a way to provide a common language for staff members and volunteers to use in describing their work.

Worksheet 2.9 can help you create or refine a program description for your technology learning center. Following the worksheet is Sample 2.1, a sample program description.

> *"A program description should answer fundamental questions about when and why the program was created, what it consists of, what its goals are, and what accomplishments it may have achieved."*

Worksheet 2.9
Develop a Program Description

Developing a clear description of your program is important in two ways. First, it will help those who work in the program remember the basics, from the hours of operation to the program's purpose and goals. Second, it will help you communicate clearly with the community and potential partners. Use this worksheet as a guide to what you might include in a program description.

Our Program Description Includes...	Included? Yes	No	Notes for Additions/Edits
Background on the sponsoring organization and mention of any major partners			
Background on the technology learning center (when, how, and why it started)			
Purpose and goals of the center			
Populations served			
Description of services and activities			
Intended outcomes for participants			
Hours and days of operation			
Other			
Major funders			
Accreditations, honors, and awards			
Special projects			

The YouthLearn Guide • Created by the Morino Institute • *www.youthlearn.org*

Sample 2.1
Program Description

Greenwood After-School Learning Center, Washington, DC

The Greenwood After-School Learning Center (GLC) serves 125 children and their families in the far northeast section of Washington, DC. The main goal of the center, sponsored by the Greater Washington Boys Clubs of America, is to provide a comprehensive, engaging youth development program for children ages 6 to 16, using technology as a primary learning tool. GLC's mission is to assist children in their intellectual, cultural, and social development by providing a coordinated set of out-of-school learning experiences guided by a corps of trained, talented instructors.

Services

GLC opened in 1986. In 1998, with the support of the Abbott Foundation, the center established a state-of-the-art technology learning center with 15 high-end computers, high-speed Internet connections, and an array of educational software, digital cameras, and other multimedia equipment. The center is open from 3:00 to 6:00 p.m. on weekdays.

The technology is not an end unto itself but only one aspect of the center—a compelling focal point for providing exciting learning opportunities in academic areas as well as in performing and visual arts, leadership, and social skill development. GLC not only provides children with understanding and knowledge about the various ways to use the technology but, perhaps more important, also infuses the technology into all content areas and employs inquiry-based and experiential approaches to instruction. After six months of regular attendance in the program, GLC children show significant gains in learning skills such as critical thinking, problem solving, self-motivation, and leadership.

The center also provides children with a one-to-one mentor program and family counseling. Summer internships are available for older teens.

Awarded the prestigious After-School Excellence certification by the U.S. Department of Education, GLC has been especially recognized for its student productions of an annual community photography exhibit and its Internet community project, which links all the local social service groups and faith-based organizations.

Step 8:
Set Up a Program Schedule

The next step in the planning process is to establish a general schedule. Although such schedules will vary greatly from center to center according to the available space, number and ages of participants, length of the program day, and so forth, some general considerations should be kept in mind as you develop your schedule, as described in the following sections.[6]

Structure and Flexibility

Ideally, your program schedule should include daily, weekly, and monthly routines as well as time for celebrations and special events. Children like to know what to expect, so it is important to have an established routine that everyone understands. Routines also allow your students to look forward to regularly scheduled events that they particularly enjoy.

At the same time, your routine must allow for a degree of flexibility. Structure is essential for making a program work, but when children are not in school, they need time to choose activities, explore their interests, hang out with their friends, and relax. A good schedule will include blocks of time that accommodate a wide range of children's needs.

Transition Time

The period when kids make the transition from their school day to their after-school activities is a time when flexibility is particularly important. When adults finish work for the day, most feel the need to unwind—some people go for a run or to the gym; others get together with friends, read, or watch television. Like adults, children and youth have different ways of unwinding. Some kids need to run around for a while to let off steam, whereas others prefer to spend some time alone or have a snack right away. Some kids just want to spend time with friends. An effective schedule recognizes those preferences and builds opportunities for different types of transitional activities into the program schedule.

The following terms provide an easy way to remember the basic types of transitional activities:

- **Laps**—participating in physical activity

- **Snacks**—having something to eat and drink

- **Raps**—socializing with friends and staff members

- **Naps**—relaxing and enjoying quiet time.

The transition time can last for 10 to 20 minutes and—depending on space constraints—can be organized so that each kind of transitional activity has a separate space, with staff members or volunteers overseeing each area. A well-structured transition time allows students to participate more fully in the activities to follow.

Other Scheduling Considerations

Along with planning the transition from the school day to after-school activities, keep in mind a few other scheduling concerns:

- **Make space for homework.** If you have a specific time set aside for students to work on their homework, make sure that students who have finished their work can participate in other quiet activities. In other words, try not to give students a choice between doing their homework and participating in an exciting new activity. Students not doing homework during homework time could be asked to read quietly until homework time is over.

- **Create structured free time.** Use at least one block of time during the afternoon to set up activity "stations." Have enough activities to ensure that no station is overcrowded, and allow students to move from one to another at their own pace. Possible activities include arts and crafts, board games, extra homework help, and movement or dance.

- **Ease the end-of-the-day transition.** Think about scheduling relatively low-key activities at the end of the program day, so that when parents pick up their children they won't have to take children away from an activity that is hard to abandon. For example, it is usually easier for a kid to leave a drawing project he or she can finish the next day than a high-energy game of capture the flag.

Scheduling and Instructors

Classes will be more successful if instructors are able to concentrate on groups made up of children of similar ages and developmental stages. Of course, organizations do not always have the staff capacity to assign a separate instructor to each age group. However, instructors who prefer working with teenagers can probably adjust their curriculum and teaching style to work with preteens. Teachers who prefer working with elementary-school-age children can probably move within the 6- to 10-year-old range.

If program circumstances require that staff members and volunteers teach across a broad age spectrum, the following strategies can help:

- Set up the program schedule to allow instructors to concentrate on similar ages or developmental groups on any given day. If an instructor has to teach teens and elementary school children in the same week, allow him to cluster the teen groups on some days and the elementary school groups on other days.

- Set up class cycles, or "rotations," for new instructors who want to learn how to teach different age groups. If an instructor loves teaching preteens but wants to try working with younger students, let the instructor complete a class cycle with the 12- to 13-year-olds before taking on a class of 7- to 8-year-olds.

- Limit inexperienced instructors to one class per cycle. It can be exciting for a staff person or volunteer who has never before taught in a structured setting to try teaching and developing curricula for the first time. It can also be exhausting and overwhelming as the new instructor learns through trial and error. Instructors who have previous experience teaching or working with children in structured settings will have a sense of their own limits and probably can decide by themselves how many classes it is reasonable to take on at once.

- Be aware of the training needs of your instructors, and whenever possible provide them with access to books, training, and other forms of staff development. It is particularly important to recognize that working with groups of children who are pre-readers or are at early stages of reading—typically ages 3 to 8—requires exposure to specialized teaching methods. The lack of such training often results in children "acting out" in classes as inexperienced instructors try to engage them in ways that do not meet their developmental needs.

Worksheet 2.10 can help you design a program schedule.

> "Your program schedule should include daily, weekly, and monthly routines as well as time for celebrations and special events."

Worksheet 2.10
Program Schedule

Use this worksheet to help you plan a program schedule. Be sure to allow time for things like homework help, transitional activities, and snacks.

Day(s) of the Week	Activity/ Age Group	Time Period	Location	Instructor and Aides	Notes

Adapted from K. E. Walter et al., Beyond the Bell: A Toolkit for Creating Effective After-School Programs, *Second Edition (Naperville, IL: North Central Regional Educational Laboratory, 2001), p. 76 (www.ncrel.org/after/bellkit.pdf).*

The YouthLearn Guide • Created by the Morino Institute • *www.youthlearn.org*

A Special Note on Creating a Positive Climate

Learning centers have climates, just as cities or towns do. It's something you feel, something in the air. It's more than how a place looks, although that's certainly important; you also sense it in how people interact, in how they listen, and in what they say. The environment of your center will determine how effective you can be as a teacher and leader of children. Is it a nice place to be? Do kids feel safe there? Do they know what's expected of them?

The Dimensions of Climate

Four factors will contribute to the climate of your technology learning center: values, environment, patterns of interaction, and people. If you get all of them working together and reinforcing one another, you'll be much more effective and the kids will be much more successful. As you plan your center, the key question to ask yourself is, What would someone walking into the center for the first time think?

Values

Values are the core concepts you want kids to learn and are the basis for how you want your center to operate. Among all the center's goals, just a few concepts really hold everything together: "Treat everyone with respect." "Our sense of community means we look out for all members of our program." "Sharing ideas and collaborating brings us better answers."

Look for ways to reinforce your core values in all aspects of your center. If having a sense of community is one of your values, try giving the children a community-building task when they come in each day. For example, each child can take his or her picture out of a basket and put it on an attendance chart. If someone is absent, the others should think, "Someone in our community is missing today. Who isn't here?"

If one of your values is collaboration, students must see you working with your peers; otherwise, you will send contradictory messages. Keep something else in mind: The rest of the world may be full of messages that work against your values, so it's even more important that you use every opportunity to reinforce the beliefs and behavior you're striving for.

Environment

Environment includes the physical aspects of your center. You can't control everything, but you can control a lot, such as how you arrange the room, what you put on the walls, and how you have children use their space. All of those factors should reflect your program's core values. The key is to be purposeful—all aspects of the environment should be based on a reason that supports the goals of your program.

For example, will you use tables or desks? Will they be arranged to encourage individual or collaborative work in teams or pairs? The walls of your center are a canvas; what will you paint on them? There's no single right answer, of course, because what you decide is based on your program's core values, but ideas for things to put on the walls include pictures of the kids, work the kids have produced, job charts, schedules, pictures of other people in the center or community, poems, proverbs that reflect the center's values, biographies, and quotes of the day.

> *"The environment of your center will determine how effective you can be as a teacher and leader of children."*

The only wrong answer is to do nothing or to do things that do not reinforce your core values. For example, why put up poems or quotes if your class won't do much work on language arts? It's not that it's a bad idea, just that you would be wasting space that could reinforce your real focus. And if you're going to put up a quote of the day, what else will you do to emphasize its message? You should at least hold a discussion about the quote each day, or you can take it even further by having different kids bring in a quote and explain why they think it's meaningful.

If your center will have tremendous demands on space, you may have to move from room to room on different days. This can go against the feelings of safety and community you're trying to instill, so prepare yourself to deal with it as well as possible. Think about making it possible to keep work and creative materials in containers that kids will find familiar and which will be easy to grab and take with you. Use techniques like carrying schedules and posters showing students' assigned tasks to the new space. Be creative in evoking a portable sense of familiarity no matter where you may be.

> " *People are what a learning center is all about.* "

Patterns of Action

Patterns of action are our ways of and expectations for interacting with one another. One of the best things you can do is to help kids understand what's expected of them so that they can feel part of a larger whole. This approach can include specific job responsibilities, consistent schedules, and ground rules (all of which can make great charts to hang on the wall, thus using the environment to reflect values that guide patterns of action).

Children often are disempowered when put in situations in which they have to wait for someone to tell them what's happening or what to do. Job charts and schedules that detail each child's task or duty in the room can be powerful tools for teaching responsibility and pride in a task well done. Patterns of action involve behavior—how kids walk in the hall, how they act in class. Part of your goal will be to guide those patterns in a positive way, so you will have to be conscious of your own patterns of action when dealing with your colleagues and the kids.

People

People are what a learning center is all about. Remember that a lot of people can help you do your job if you'll only use them. From the security guard to the groundskeeper, from parents to people in the neighborhood, everyone can be a positive influence. If, for example, one of your center's goals is to be an integral part of the larger community, think about what you will do to demonstrate that to children. Will you post pictures of people and places in the community? Will you have guest speakers? Will community members participate in your projects?

Worksheet 2.11 can help your planning team explore ways of creating a positive learning climate in your technology learning center.

> @ *For more information, examples, and resources, see the extensive YouthLearn website at www.youthlearn.org*

Worksheet 2.11
Create a Positive Climate

List four or five "core values" for your technology learning center. Core values are broad statements (reflected in the vision statement for the center) that concisely describe the beliefs that are fundamental to your learning center.

1.
2.
3.
4.
5.

In what ways will your center's environment reflect and reinforce those core values?

Core value	Ways in which the environment will support that value
1.	
2.	
3.	
4.	
5.	

In what ways will you instill "patterns of action" that reinforce those core values?

Core value	Patterns of action that will support that value
1.	
2.	
3.	
4.	
5.	

In what ways will your center's interactions with people (staff, children, and community) reflect its core values?

Core value	Ways in which interactions will support that value
1.	
2.	
3.	
4.	
5.	

chapter 3

MANAGE YOUR PROGRAM

Overview

Now that you have created the core of your program, how do you get it off the ground? Making decisions about how to launch your program, how to run it on a daily basis, and how to ensure its quality is crucial to its long-term success.

This chapter describes ways to accomplish the following important tasks:

- Get the word out in the community about your program

- Register children in your program

- Identify policies and procedures you need to have in place before you open your doors

- Create a budget

- Find other organizations in your community to join forces with in order to expand your impact in the community

- Develop a process of continuous improvement for your program.

Contents

- Basic steps for managing your program on a daily basis
- Strategies for engaging the community
- Guidelines for registration and tracking
- Worksheets for effective management

Audience

- Program directors
- Organizational leaders

Outreach and Communications

Even the best-planned program will not succeed without proper outreach into the community it serves. Call it outreach and communications, marketing, or public relations, but in establishing a youth technology program, as in running any kind of business, you will need to get the word out in your community. When and how should you reach out, and to whom?

When Should You Start?

The most successful outreach efforts will begin early in the life of your program, before its doors open to participants. If done correctly, outreach and communications will help establish good ties with the community and bring in the kids for whom you are creating the program. Finding kids to fill your roster is not a given; you must build a positive relationship with the larger community so that parents, educators, local businesses, and other nonprofits will feel comfortable supporting your program.

Regardless of when you start, you should have a solid strategy for how you will conduct and keep track of your outreach efforts.

Who Should Be the Target of Your Outreach Efforts?

Focus your outreach efforts on the following groups of people:

- Potential program participants

- Parents

- Teachers

- Local businesses

- Potential volunteers and staff.

Possible Strategies

A variety of strategies can help you build the relationship between your program and the community surrounding it:

- **Newsletters and brochures.** Communicating with people need not be sophisticated; glossy brochures are nice if you have the money or if someone offers to donate services, but simple flyers can be just as effective—and sometimes more so. Think about your audience: If you are trying to attract people who don't have a lot of money, expensive paper and a fancy layout might convey the idea that your program is expensive and out of reach.

- **Local newspapers, radio, and television stations.** Most local media produce human-interest segments about events in the community. See if you can arrange for them to cover the opening of your program or to come in and cover an average day once the program is up and running.

- **Community bulletin boards.** Here you can post the job openings and volunteer positions available in your program or publicize an upcoming event.

- **Promotional items.** If you are creative, you can come up with lots of low-cost giveaway items that promote your program. Local businesses might be willing to help with the costs of producing calendars, magnets, T-shirts, posters, and other items with your program's logo and contact information.

- **Conversations at local gathering places.** Many grocery stores will let community organizations set up tables for outreach and educational purposes. Block parties, community markets, and local fairs are other good places to talk in person with community members and spread the word about your program.

The key to building strong ties with your program's community is consistent contact—keep following up, and make sure your audiences know that you care about what they have to say. Use everyone you know in the community for ideas and connections. If you want to print T-shirts for your program, for example, ask people if they know of a business that could help. Keep asking and keep communicating!

Worksheet 3.1 will help you develop an outreach and communications plan.

" The key to building strong ties with your program's community is consistent contact."

Worksheet 3.1
Outreach and Communications

Reaching out to the community will help you build a broad network of support among participants, partners, qualified staff, and volunteers. Use the questions in this worksheet to help you refine an outreach and communications strategy.

Questions to Ask About the Outreach Strategy

☐ Who are our audiences? Have we identified them all?

☐ Have we targeted all messages appropriately to each audience?

☐ How will we follow the progress of our outreach strategy?

☐ Do we maintain an ongoing record of communication with each audience?

☐ Do we regularly review our outreach strategy to make sure it is still meeting the needs of our audiences?

☐ Have we given one person or group of people associated with our program ultimate responsibility for communications?

Questions to Ask About Each Audience

☐ What kinds of information about this program does the audience want or need to have?

☐ How does the audience like to receive its information? In writing? In person?

☐ Does the audience have a leader who influences the group's opinions?

☐ What is the level of trust between the audience and the program?

☐ How can we build that trust? Who is the best contact in our program to interact with this audience?

Questions to Ask About the Community As a Whole

☐ What percentage of the community has school-age children?

☐ What are the primary languages in the community?

☐ Who are the primary employers in the community?

☐ What are the largest and most active faith groups in the community?

☐ What are the major sources of news and information in the community?

Questions to Ask When Conveying a Message to the Community

☐ What do we hope to achieve with this message?

☐ How will the recipients of this message feel about it?

☐ What language should be used for this message, given its audience? Is it written at the appropriate reading level and in clear language?

☐ Have we requested feedback so that we can learn how to communicate better in the future?

Adapted from K. E. Walter et al., Beyond the Bell: A Toolkit for Creating Effective After-School Programs, Second Edition (Naperville, IL: North Central Regional Educational Laboratory, 2001), pp. 147-48 (www.ncrel.org/after/bellkit.pdf).

Registration and Tracking of Participants

Registration

The registration of all participants in your technology learning center is one of the most important aspects of running an effective, goal-oriented program. Registration establishes commitment on the part of the children and families involved, ensures clarity about the nature and activities of the program you are offering, and sets expectations on both sides.

The process of registration begins with a clear communication of the program's purpose, a program description and schedule, entry requirements, expectations of the parents' involvement, fees (if any), and conditions for continued participation. It should include a face-to-face meeting with the children register- ing, along with their parents or guardians, to ensure a common understanding of parental and participant expectations as well as of the program guidelines. By the end of the meeting, everyone should have copies of completed registration forms and a simple personal plan with a schedule and goals.

Tracking

Tracking—following the progress of every participant—is the best way to support the personal growth and learning opportunities available through your technology learning center. Tracking offers you timely information so that you can provide (or refer children to) appropriate physical, social, educational, and health services. Moreover, it opens an impor- tant door to engaging parents, families, local service groups, and volunteer mentors, who represent additional caring and nurturing adults for the young people in your program.

Methods of tracking cover a wide range of practices, from staff sessions that focus on the progress of individual children to customized databases that host all the information about any given child or family, including attendance, progress and growth, resources available to meet specific needs, and communication records.

Worksheet 3.2 can help you create a plan for registration and tracking activities.

Worksheet 3.2
Registration and Tracking

Use this worksheet as a starting point for developing and maintaining a registration and tracking process. Note on the sheet which staff member is handling which tasks, and by what date.

Action	Person in Charge	Due Date	Status	Issues
Develop a registration process for your technology learning center.				
In creating a registration process, be sure to compile the following information: • A description of each class and activity • A description of intended field trips and community activities • The maximum number of students the program can accommodate • A timeline of the program by the semester or year. The registration process should include: • A meeting between parents and key staff in which personal goals are set for the student (see below) • Receipt of a completed, signed registration form from the parents • Clarification of expectations for parents and students and their commitment to participation.				
Produce a registration form.				
The registration form should capture the following information: • Demographic information (age, gender, etc.) • Parent or guardian name and contact information • Emergency contact/pick-up information • School information • Information on eligibility for subsidized meals • Subsidy or payment schedule agreement • Consent for medical and behavioral testing, field trips, and other activities for which parental consent is required.				
Meet with each student and his or her parents or guardians.				
Family meetings offer an opportunity to discuss program information and goals with parents or guardians and children. Be sure to cover the following topics: • Parents' expectations and goals for their child • Student's own interests and goals • Student's learning strengths and challenges • Student's social strengths and challenges • Agreed-upon goals and program schedule.				

(continued on next page)

Manage Your Program

3

Worksheet 3.2 continued

Registration and Tracking

Action	Person in Charge	Due Date	Status	Issues
Develop a tracking system for student attendance, participation, and accomplishments.				
A tracking system helps you monitor both your students' progress and progress toward your goals. Developing the system should include the following steps: • Investigate options for automated student logs and tracking systems. • Decide on the tracking system to be implemented. • Implement the tracking system. Create a log for staff members to use in tracking each child. Tracking logs should include information on: • Attendance • Participation • Noteworthy achievements • Problems and challenges.				
Develop session and class schedules.				
• How many students will fit in each class? • What factors will be used in deciding how to place students in appropriate classes (age, gender, etc.)? • How many times a week should each class meet, and for how long? • What are students' present schedules and conflicts? • Who will create a master written class schedule?				
Develop a strategy for communicating with students and parents.				
• Who will be in charge of developing and maintaining good communications with your program's students and their parents? • How will students and parents be notified of session and class schedules? Of changes in the schedules? • How will agreement be obtained from parents regarding the program's schedule, rules, and procedures?				

The YouthLearn Guide • *www.youthlearn.org* • Created by the Morino Institute.

Policies and Procedures

Before your center opens, you need to decide on the policies and procedures that staff members and participants should follow. When you are writing policies and procedures, keep your audience in mind. If you have relatively inexperienced staff members or volunteers, you will want to spell out basic guidelines more carefully. Rules for the kids in your program should be clear and direct, but don't overdo it: Too many rules will create a negative environment.

It is a good idea to consult with a lawyer or other professional for advice about what policies and procedures you need to put in place immediately. Some policies and procedures will be more important than others. For example, it is essential to establish clear guidelines for protecting the confidentiality of the tracking information you collect on each person and family involved in your program. See if someone in the community would be willing to provide the advice you need free of charge. As time goes on, you will probably think of new policies and want to change or eliminate some old ones, so keep careful track of what is current.

After you have formulated all the policies and procedures you think you will need, distribute them in writing to the staff and, where appropriate, to the parents of the kids in your program. Also post them in common areas as necessary. Your program will run more smoothly if staff members, volunteers, parents, and kids all know what is expected of them.

Worksheet 3.3 is a list of the kinds of questions you should ask yourself when creating policies and procedures for your center.

> *"Your program will run more smoothly if staff members, volunteers, parents, and kids all know what is expected of them."*

Worksheet 3.3
Policies and Procedures

Use this worksheet to help you determine what kinds of policies and procedures you need to consider before you open your center's doors. (Be sure to investigate which ones are required by law in your area.) If you reach out to members of your community, you may be able to find legal professionals who can help you draft the most important of these policies at little or no expense to you.

General

☐ Who can be admitted to the program? Are there restrictions based on area of residence, age, etc.?

☐ What is the maximum number of participants we can handle?

☐ What are the hours of operation?

☐ On which holidays will the program be closed?

☐ What is the procedure in case of bad weather?

☐ What is the policy if a parent does not arrive to pick up his or her child?

☐ What are the general transportation policies (if any)?

Technology

☐ Is there an acceptable-use policy for the computers in our center? Have restrictions been clearly laid out (e.g., a ban on copying center software, on bringing in outside software, or on intentionally sending viruses)? [You should include the right to refuse access to anyone who violates any of the rules or procedures.]

☐ Has appropriate use of the Internet and email been clearly stated? [For example, you might want to state that it may not be used for reading or distributing pornography or other objectionable material or for sending threatening messages.]

☐ What is the procedure if a program participant violates the acceptable-use policy?

☐ Is all the equipment fully covered by insurance?

☐ What is the daily/weekly/monthly routine for keeping the equipment clean so it will function properly?

☐ What is the daily/weekly/monthly routine for backing up files?

☐ What is the procedure when a piece of equipment breaks or malfunctions?

☐ Who is responsible for closing down the computers and locking up at the end of the day?

(continued on next page)

Staff/Volunteers

◻ What are our hiring policies? When should we conduct background checks?

◻ How and when will staff and volunteer evaluations be conducted?

◻ What will be done in the event of unsatisfactory performance by a staff member or volunteer?

◻ Under what conditions will employment be terminated?

◻ What are the provisions for staff development?

Program Participants

◻ What are the basic rules for the program (e.g., restrictions on bringing in food and drinks, sign-in procedures)?

◻ How will children with special needs (e.g., food allergies or physical limitations) be accommodated?

◻ Whom should participants talk to if they have a problem with a staff member, a volunteer, or another participant?

◻ How should a physical injury or accident be handled?

◻ What should be done if a program participant is found to have drugs or a weapon?

◻ How will inappropriate behavior or acceptable-use violations be dealt with?

◻ What should we do if we suspect that someone in our program is being abused? [Note: Check your local laws to find out what you are legally required to do in those situations.]

◻ How will common disciplinary issues be handled? Which behaviors will result in suspension or expulsion from the program? [Think about how specific you want to be.]

Safety/Legal Issues

◻ Does the program meet state licensing requirements?

◻ Does the program have liability insurance to cover any possible lawsuits (e.g., if someone is injured)?

◻ What is the procedure in case of fire and other emergency situations?

◻ How will program participants with medical conditions or who require medication be handled?

◻ Have we provided a general release for participants' parents to sign? [This will enable you to use the names and photographs of participants for promotional purposes, such as in brochures or on the web.]

◻ Is there a standard field trip permission slip? Are there general policies about where and when field trips can be held and how much advance notice is required?

Budgets and Reports

It is important to consider financial issues early and carefully, because they will affect almost all the decisions you make about your program. In creating a budget, be as specific as possible. Include start-up costs as well as the operating costs you anticipate incurring over the course of the first year.

As with creating policies and procedures, you may need some outside help with this task, so see if any accountants or bookkeepers in the community might be willing to provide you with some free advice. In addition, you may want to ask people who run established out-of-school programs in your area how they developed their budgets and what surprised them during the first year.

As you develop your budget, examine the sources of income (e.g., grants, donations, and fees) that will fund your program. Funders these days are especially interested in detailed grant reporting. They are likely to ask exactly how much you spent, what you spent it on, and what results you achieved. Be sure to keep good records—it could affect your ability to maintain (or increase) funding in future years.

Budgeting is a vital part of any successful program, and it should be refined every year. After you begin operations, take careful notes on expenses that you hadn't expected, items for which you over- or under-budgeted, and other differences between actual and projected costs. This will help you improve the accuracy of your budgeting in subsequent years.

Worksheet 3.4 lists some categories of expenditures to help you create your budget.

✔ Reality Check

Quality learning programs aren't cheap. You'll need money for supplies, reusable equipment, and field trip transportation and entrance fees. Additionally, a lot of staff time is needed to plan projects, acquire supplies, coordinate field trips, and carry out project-related activities. When evaluating whether your program is financially ready to proceed, you might ask yourself the following questions, among others:

☐ Does the program budget need to be expanded or reallocated to make increased funds available for staff time, supplies, and other essential resources?

☐ Has money been allocated for staff training?

☐ Is the process for purchasing project supplies straightforward and fast enough to enable staff members to acquire what they need in a reasonable time frame?

☐ Can a reimbursement system or petty cash fund be set up to enable staff to make smaller expenditures in a very short time frame if necessary?

☐ Have reliable partnerships or donation sources been set up to help meet resource needs?

Worksheet 3.4
Budget Development

It is important to get a complete financial picture of your program before you begin to spend any money. Use this list to help you establish the initial set-up costs and the ongoing operating costs of your program.

Set-up Costs

☐ Costs to build or modify the space for your program

☐ Costs of equipment or software you were not able to have donated to your program

☐ Costs of staff recruitment, such as ad-placement fees and any initial training

☐ Publicity costs, such as promotional flyers and neighborhood information sessions

☐ Legal and accounting fees associated with setting up policies and procedures

☐ Legal and accounting fees associated with creating a budget and an expense-tracking system

☐ Licensing and incorporation costs

Operating Costs

General

☐ Rent and utilities

☐ Salaries and benefits for all staff

☐ Custodial costs

☐ Transportation services to and from the program (if applicable)

☐ Insurance premiums for coverage of property, contents, and liability

Technology

☐ Equipment costs are considerably higher for technology-based programs than for traditional out-of-school programs. Once you have your equipment, some of which might be donated, you must have enough in your budget to maintain and care for it.

☐ Internet connections are vital; get the fastest, most reliable connection you can afford.

☐ Software is an ongoing cost. You will need to keep it updated, and you will probably want to leave room in the budget for new software.

(continued on next page)

(Worksheet 3.4 continued)

Budget Development

Operating Costs *continued*

☐ Furniture should accommodate people of all sizes and shapes; chairs in particular should be adjustable so that all participants, regardless of height, can be at the right distance from the monitors and keyboards.

☐ Security, such as an alarm system, should be included in your budget.

☐ Peripherals are necessary. Remember that each computer station needs at least a keyboard and a mouse, which are often not included in donation packages.

☐ Technical support is critical to the success of a technology-based program, and it is vital to have good help. Volunteers are a good idea in theory, but it can sometimes be difficult for them to assist you on short notice. If you want immediate assistance, you will probably have to pay for it.

Supplies

☐ Creative materials, such as pens, paper, glue, scissors, index cards, and books, are vital to your program.

☐ Printing supplies, such as paper and ink cartridges, are expensive and add up quickly.

☐ Newsletters and other forms of communication can involve printing costs and postage.

Miscellaneous

☐ Food for snacks should be included in your budget.

☐ Field trip expenses, such as transportation, should be given a generous allowance.

The YouthLearn Guide • Created by the Morino Institute • *www.youthlearn.org*

Partners

Remember that your center is not just a building with staff and kids; it is part of a much larger community where all kinds of people are willing and able to help you reach your goals. Partnerships can be complicated, but they are worth the extra work if they expand the opportunities you are able to offer the kids in your program.

What Is a Partnership?

A partnership is simply a relationship between your program and another entity in the community. It can take many forms, but ideally it will be a true collaboration, not just a simple sponsorship. A collaborative relationship is ongoing and benefits your organization as well as your partner's. Although it can be a sticking point, sharing leadership can create a powerful positive force in the community, far greater than each organization could exert on its own.

When Should You Start Thinking About Partnering?

Begin to think about partnerships as soon as you have identified your program's mission and scope. Partners might be able to help get your center up and running with donations of equipment, furniture, expertise, and time. You should, however, have a solid plan in place—valuable partners will not want to waste their time on a program they don't think is going to get off the ground.

How Do You Set Up Successful Partnerships?

Your students, their parents, and your staff are a great resource for discovering who in the community might be interested in partnering with your center. Ask them if they will help you brainstorm about potential partners in the community. That way, you can also make the most of personal connections—for example, a parent might belong to a church that is looking for a project with kids in the community. Ask them to ask the same questions of other people they know, and begin creating a network of people who are talking about how to help your center.

When you are considering a partner, look for the following characteristics:

- Shared beliefs or a similar mission

- A willingness to work together to achieve your goals

- Open and honest communication

- Mutual gain—is everyone gaining something from the partnership?

Who Are Potential Partners?

Many resources in your community could serve as valuable partners:

- Parents of the children in your center may be available as volunteers or be willing to help with scheduling. A parent might be willing to create or write for a newsletter for your center.

- Other nonprofits and community organizations can be approached about sharing space or pooling resources to share the costs of a staff development program.

- Churches, synagogues, and other faith-based organizations might be able to provide you with potential participants for your center, once it is up and running.

- Local businesses are often willing to make donations of money, equipment, furniture, or services.

- Area colleges and universities sometimes offer the services of their students as a way of connecting with the community and giving their students real-world experience. Universities are great places to look if your funders require a detailed evaluation of your program; graduate students frequently seek such projects.

- Libraries are filled with learning resources and are generally staffed with people who want to help kids learn.

- Senior citizens centers might offer older people as mentors to the young people in your program.

- Police departments are frequently involved in public service projects and may be able to help you get the word out about your center.

In short, the kinds of organizations your center can partner with and the types of partnerships you can create are limited only by your imagination. The key is to get the word out—and keep asking!

How Do You Evaluate Your Partnerships?

As your partnerships progress, it is important to stop and ask yourself the following questions:

- Is this partnership working?

- How can the partnership be improved?

- Are there other partnerships we should be pursuing?

Meaningful partnerships can be time-consuming and challenging to maintain, but they are an excellent way for an individual organization to have a greater impact on the community than it would have on its own. Whether it involves pooling resources, sharing facilities, or combining efforts in some other way, creating partnerships makes it easier to make a difference.

Continuous Improvement Through Evaluation

Continuous improvement should be a primary concern of every technology learning center. Continuous improvement is the process by which you regularly, systematically, and honestly evaluate what your center is doing well and what it needs to improve. Today, more and more funders are requiring this kind of comprehensive evaluation as a condition of continued financial support.

Early in the process of establishing your program, you should begin to set up a system of ongoing evaluation. A strong evaluation design will provide information that you can use daily to fine-tune your program.

To understand the continuous improvement process, it is important to understand the terms of evaluation. Evaluation is the process of analyzing data to assess what works and what does not work in achieving goals. In this context, the word "goals" is synonymous with "objectives,"

"outcomes," and other terms your program may use to convey changes you expect or desire.

Data collection is only a part of full evaluation—one step among many. The evaluation forms that staff and participants are asked to complete are *tools* for data collection; they are not the evaluation itself. To be useful, data collection has to be followed immediately by analysis. This analysis drives the continuous improvement process.

Outcome Measurement

Currently, most funders prefer an evaluation process called "outcome measurement," because they see it as the best gauge of how a program is meeting its objectives. Traditional models of evaluation focused on outputs, such as products delivered (e.g., classes held) and people served. Outcome measurement goes further: It takes outputs into account, but it focuses on how well these outputs produce end results (e.g., measurable improvements in reading ability).

Outcome measurement is not easy. It can be expensive and time-consuming, especially if it is not introduced into the initial design of the program. It requires you to establish appropriate and reasonable desired outcomes, which can be a complicated process. And if a program is not achieving its outcomes, outcome measurement doesn't always indicate exactly *why*.

On the positive side, however, outcome measurement can help with the following tasks:[7]

- Help an organization clarify its mission and ask, "What exactly are we trying to accomplish, and how are we going to accomplish it?"

- Increase staff and volunteer satisfaction by providing evidence, when appropriate, that the program is meeting its goals

- Provide a steady stream of information and feedback to help with the continuous improvement process

- Identify services that need more attention

- Assist in developing and justifying program budgets

- Display a program's achievements for board members

- Help define a program's long-term goals.

In short, outcome measurement is a big step beyond simple data collection. As a result, it is a considerably more involved process and yields considerably more valuable results. Given that today's funders are demanding more accountability from the programs they fund and that government agencies and national nonprofit organizations are moving toward outcome measurement as their primary method of evaluation, there is great incentive to explore it early in the process of planning your center.

Who Should Do the Evaluation?

As you begin to examine what system of evaluation is right for your center, you should answer the following questions:[8]

- Do we have the expertise on staff to design and conduct the evaluation?

- Do we have the time to manage all aspects of the evaluation?

- Are we able to look objectively at our work?

- Will our funders permit us to do our own evaluation?

If you answer no to any of those questions, you might want to consider hiring an external evaluator or team of evaluators. An evaluator can be either someone who holds an advanced degree in evaluation or an experienced practitioner in your content area. Colleges and universities are a good source for external evaluators. There are private firms that specialize in program evaluation. It is also possible that your school district has people who are experienced in conducting evaluations. Whenever possible, work with an evaluator who is familiar with your type of program.

Emphasize the Ongoing Nature of Evaluation

An annual evaluation may allow problems to remain for an entire year before they become apparent and steps are taken to address them. To avoid this unnecessary delay, programs should make time and resource decisions that allow for ongoing assessment. Consider the following ways to continually integrate evaluation:

- Provide structured time for program staff to collect important data and to meet and discuss the long-term goals of your program.

- Make the review of evaluation results a regular part of meetings between program partners.

- Establish a procedure by which changes will take place. Determine the "who, what, when, where, and how" for distributing evaluation results and other information. Gather suggestions for changes. Decide which ones to follow through on, and then make the changes.

- Develop a strategy for keeping all potential supporters and participants informed about the process of change. All people who have a stake in the program will need to be notified of the evaluation results and informed about the process by which changes based on those results will be made.

- Promote a culture in which the purpose of evaluation is to create continuous improvement. Although this may be easier said than done, programs can try to foster a positive attitude toward evaluation by celebrating the opportunity for improvement rather than bemoaning it as another bureaucratic requirement. Strong, positive leadership makes a difference in staff perceptions of evaluation activities.

- Bring partners together regularly to discuss data in light of "big picture" issues. The daily demands of running a program lead to a "putting out fires" management approach. The process of data collection and analysis offers the opportunity to refocus on both collaboration and the underlying purposes of operating a program.

@ *For more information, examples, and resources, see the extensive YouthLearn website at www.youthlearn.org*

Overview

Having an excellent staff is the key to the success of any learning program for youth. All the software, hardware, lesson plans, and art supplies in the world are useless without dedicated, trained, fulfilled, and caring staff members.

The first step to building an effective team for a technology learning center is rigorous recruitment and evaluation of candidates. Getting the word out to the right candidates and appraising them thoroughly and carefully will help you find the best staff members for your program.

What Positions Should You Create?

Budgets are often tight, and staff members in youth development programs sometimes have to wear multiple hats. Start with planning for a "dream team," and then consider budgetary limitations.

Your ideal staff might include the following positions:

- Director of youth development

- Director of technology learning programs

- Instructors. Recommended ratio: 1 instructor for every 15 children. (Some can be part-time if necessary.)

- Assistant instructors or teaching aides. (These can be teens or college students, but you shouldn't use them unless they receive appropriate training and are always supervised.)

You may also want to have a curriculum developer (part- or full-time) and a community outreach worker.

A center director and instructors sometimes may need to pinch-hit on technical support and maintenance, but ideally your program will have a part- or full-time network administrator who provides technical support. This person can also sometimes serve as a web developer for your organization.

Although the program staff may contribute to fundraising and organizational administration, other staff in the sponsoring organization should have the primary responsibility for such tasks.

Contents

- Basics of identifying staff needs, recruitment, and appraisal of applicants

Audience

- Program directors
- Organizational leaders
- Educational administrators

If children have not mastered the basic skills of reading and writing, technology alone will not do the trick. Before your program acquires computers and software, staff members must consider how to provide motivation and experiences that will help children improve their reading and writing. You might ask the following questions:

☐ Do staff members need training in effective strategies for improving children's reading skills, such as how to select age-appropriate books and how to lead read-alouds?

☐ Do staff members need training in how to lead writing activities, such as making journal entries and vocabulary building?

☐ Do staff members need additional education to improve their own writing skills?

☐ Has program time been set aside for daily writing activities?

If there is a problem in any of these areas, be sure to address the issue before you embark on a technology investment.

Recruitment: Find the Best Staff

Before you spread the word about the positions you want to fill, you need to define the roles and responsibilities of each position carefully. A detailed job description will help candidates understand the skills, experience, and characteristics the position requires. (See the sample job descriptions at the end of the chapter for director of youth development and director of technology learning programs.)

Before you post your job listing, make sure that it contains an email address for responses (one that you check often) as well as a website address if your organization has one. A surprising number of organizations distributing announcements for Internet-related positions do not provide an email address to which

candidates can reply. Experienced candidates will not apply to such organizations; an organization's use of email and the web is an indicator of whether it is "walking the talk" of what it is trying to do with youth and technology. Similarly, candidates who cannot email a resume or send web-based information upon request may not have the appropriate skills and experience for the position. If a candidate has thoroughly reviewed your organization's website, it's a quick indication that he or she has basic web surfing skills, is truly interested in the position, and has taken time to prepare for the interview.

How do you find the right people for your program? People who know your organization and what type of help you need are the best sources for staff candidates. Below are a few tips for how to find great candidates for open positions in your center:

- **Talk to people who are doing well in the kind of job you are trying to fill.** Some of them might be interested in a new employment opportunity, or at least know of others who are. Ask them how they got their jobs and where they think you should post your announcement.

- **Check with organizations that know the type of people you want to hire.** This group includes funders, advocacy groups, consultants, and direct service organizations.

- **List your opening in different venues.** Use both traditional and electronic media to distribute your announcement, including organization bulletin boards (both traditional and electronic), electronic mailing lists, websites, and print publications.

- **Go online.** The Internet is probably one of the most likely places to locate candidates with technical skills and knowledge. Electronic mailing lists will usually carry job announcements for free. Some websites ask for a fee to post an announcement for a certain length of time.

- **Advertise in newsletters.** If your organization has a newsletter or other regular publication, whether hard-copy or electronic, announce your opening in your own materials. Ask others to post your announcement in their newsletters.

- **Use your newspaper.** Consider advertising in the local newspaper.

- **Ask at educational institutions.** Remember to contact your local universities and community colleges—the perfect candidate might be finishing a course of study that suits your needs. Many universities have begun to charge for posting announcements, but check to see if you can post an announcement in the career placement office for free.

- **Contact professional technology organizations.** Many areas have professional organizations for people who work in the information and communication technology fields—webmasters, network administrators, web content producers, and related professionals. Many such organizations have members-only job boards and electronic mailing lists and would be willing to share your announcement with their members.

The Selection Process

After you have gathered the names of a number of candidates, you can start narrowing the pool. The successful candidate for each position should in most cases undergo two or more rounds of interviews. Here are some suggestions for moving through the process efficiently:

1. **Conduct a telephone interview.** Unless the candidate is someone known and recommended through a trusted source, conducting the first interview over the telephone is a good idea. A preliminary telephone conversation will allow you to gauge a candidate's understanding of the position and his or her level of interest; it also provides an opportunity for initial questions and answers. You most likely will be able to weed out several candidates following telephone interviews, thereby saving time and staff resources.

2. **Arrange a face-to-face interview.** Think carefully about who on your staff should meet with each candidate. In addition to supervisors, it is usually a good idea to include at least one staff member who would be working with that candidate at a peer level. Staff members in similar positions know best what the job you are hir-

ing for entails. It is also important that they support the decision to hire the candidate because they will need to work closely with that person. Prepare an interview evaluation chart (see Worksheet 4.1) to help you and your colleagues structure the interview and the subsequent evaluation.

3. **Ask candidates at the final stages of consideration to complete a self-assessment.** (See Worksheet 4.2.) Doing so will help you determine each candidate's strengths as well as areas requiring support and development. It can also provide useful information about a candidate's general attitude and self-perception.

4. **Check references!** It is surprising how many people do not check a candidate's references. A former employer can provide insight into a candidate's work habits, knowledge, and skills as well as offer information about how quickly the candidate picks up new skills and his or her general attitude. Be specific and ask plenty of questions. Often a former employer is reluctant to make criticisms but will answer specific questions honestly when asked.

5. **Do a background check.** It is unpleasant to think about, but people who prey on children frequently apply for jobs that give them direct access to kids. Legal issues regarding background checks vary from state to state, so check with your local police department to find out what is required, suggested, and forbidden by law in your area.

The process may seem like a lot to undertake, but it is well worth the effort to find the best people you can for your staff. No other decision you make will have as big an impact on your program.

Prepare for the Interview

Before you interview candidates, you'll want to prepare by considering the following questions and related suggestions:

What characteristics are you looking for in a candidate?

- Genuine passion and calling to work with children

- Flexible: able to respond to rapid change

- Hard worker: willing to go above and beyond minimum effort

- Hands-on attitude: willing to do whatever needs to be done

- Team player: willing to share information, communicate, give and receive feedback

- Learner: eager to advance personal development and improve performance through taking responsibility and exercising initiative to learn new things; participates actively in training and other developmental experiences

- Confident: willing to take risks and to lead and model for others.

What skills and experience are you looking for?

- Strong oral and written communication skills

- Experience working with children in a program with defined goals and expectations

- High comfort level and competency with standard software and Internet applications

- Understanding of how the Internet and related technologies can be used to support learning

- Understanding that the learning program will teach more than technical skills.

What other questions should you ask the candidates?

- "Can you give me an example of a time when you felt really great about achieving an important goal as you worked with children?"

- "Can you describe an instance when you had to adapt to something that suddenly changed in your work environment?"

- "Can you give me an example of a time in the past when you took initiative in a work situation?"

The above questions are by no means the only questions you should ask; they are meant to be starting points to get you thinking about what you would like to know about a potential staff member.

For each interview, prepare an interview evaluation chart (see Worksheet 4.1); doing so will help you focus your questions, make sure you have asked all the critical questions, and organize the information for evaluating the candidate. During the interview, be sure to keep the questions you ask focused on the job.

Worksheet 4.2 is a self-assessment chart that you can modify and ask candidates to fill out. The chart also can be used to give your present staff members an opportunity to think about and assess their skill level and identify areas where they want or need additional training.

> @ *For more information, examples, and resources, see the extensive YouthLearn website at www.youthlearn.org*

Worksheet 4.1
Interview Evaluation Chart

Use this worksheet as a starting point for evaluating candidates for staff positions. If there is more than one interviewer, the chart can be used as a way for interviewers to compare their assessments of a candidate. Be sure to adapt this worksheet to include any special skills you require.

Interview Date

Candidate Name

Interviewers

Area	Weak 1	Satisfactory 2	Strong 3	Comments
Communication skills				
• Oral • Written sample				
Teaching experience				
• Experience with children • Outcomes-based work • School or community work • Has taught children • Has designed projects or curricula				
Cultural fit				
• Work ethic • Flexibility • Collaborative work style				
Experience with the Internet and related technologies				
• Software • Hardware • Web • Email • Multimedia				
Knowledge of our organization and program goals				
Other				

Worksheet 4.2
Self-Assessment Tool

Please provide an honest assessment of your skill level and experience in the following areas.

	Skill Area	Experienced/would do well	Have had comparable experience	Could learn fairly easily/ would like to learn	Wouldn't be good at it	Comments
1	Invent or organize learning activities for young children					
2	Hold the attention of a group of young children					
3	Offer constructive critique to co-workers					
4	Remain calm and take responsibility in a crisis situation					
5	Teach conflict resolution to children					
6	Work effectively as part of a lab team					
7	Show co-workers how to use email					
8	Is there any skill you would like to mention that has not been covered by the above?					
9	Please list the computer systems and software with which you have worked.					

(continued on next page)

 The YouthLearn Guide • Created by the Morino Institute • *www.youthlearn.org*

Staff Your Program

4

10	Do you have any experience with the following skills? (If yes, please explain in the box on the right.)	
	Website development	
	Use of email for educational projects	
	Photography	
	Video production	
	Desktop publishing	
	Drawing, graphic design, or other visual arts	
	Music or performance arts (dance, theater)	
	Creative or journalistic writing	

⭐ **Sample 4.1**
Job Description (Director of Youth Development)

Position Title: Director of Youth Development	
Organization Description	[Insert a one-paragraph description of your organization and its technology learning center.]
Position Summary	[Organization name] seeks an experienced, competent, and caring person to develop and manage a comprehensive youth development program for school-age youth. The director of youth development is responsible for coordinating the development and implementation of a [insert your program type: after-school, out-of-school, etc.] youth program that integrates the resources of a technology learning center into the program activities.
Responsibilities	• Work with executive leadership and staff to design a cohesive and comprehensive youth development program for youth ages 5 through 18, using the resources of [organization name] and the technology learning center
	• Develop and implement planning and project management processes to ensure the effective coordination and integration of the programs and resources of the learning center
	• Provide support to executive leadership and staff in the areas of planning, management, program execution, and communications
	• Provide the director of the technology learning center and other program staff with resources and day-to-day supervisory support to enable them to implement learning programs
	• Collaborate with other members of the management team to ensure the effective coordination of program schedules and related events for youth and families
	• Manage the development and maintenance of the learning center, including hardware and software purchases, installation and support, instructional materials, and physical plant
	• Assist with fundraising for educational programs and youth programs
	• Foster relationships among local schools, universities, businesses, and other community youth and education programs to enhance the development of the learning center

(continued on next page)

Job Description (Director of Youth Development)

Qualifications	• Leadership and management experience in a youth services, family support, or K-12 educational setting • Ability to articulate a vision about preparing youth for the future—from social, economic, cultural, and personal perspectives—to partners, funders, and the community • Highly positive and enthusiastic style; capable of motivating others • Skills and energies to build a team and lead effective staff development and training
Skills and Experience	• Relationship management skills and experience in fostering a team approach to youth development and creating collaboration among partner organizations in youth development • Excellent project management and planning skills • Excellent written and oral communication skills • At least three years of experience in an administrative or managerial capacity in an organization that provides direct services to youth • Experience and competencies in working in a multiracial, multicultural environment • Experience with or understanding of the development of educational programs in nonprofit or K-12 settings • Experience with or understanding of the application of information technology to educational programs for youth in nonprofit or K-12 settings
Salary	[Fill in salary information.]
How to Apply	[Fill in application information, including an email address and website.]

Sample 4.2
Job Description (Director of Technology Learning Programs)

Position Title: Director of Technology Learning Programs	
Organization Description	[Provide a one-paragraph description of your organization and its technology learning center.]
Position Summary	The director of technology learning programs is responsible for the development and ongoing operation of the learning center's educational programs and facilities.
Responsibilities	1. Manage the design and implementation of structured educational programs that incorporate technology into learning, including: • Developing lesson and project plans • Acquiring and preparing instructional materials • Recruiting participants and placing them into class groups • Scheduling classes • Communicating with parents and families (phone, written communications, and in person) • Tracking attendance • Documenting class activities and projects • Managing staff to ensure effective development and coordination of all of the above. 2. Serve as lead instructor for school-age children's technology-enriched educational programs, including: • Teaching classes • Troubleshooting software and hardware problems during classes • Working with staff to handle interventions to address children's medical, behavioral, or family crisis issues that arise during classes. 3. Work with the director of youth development to manage technical development and maintenance of the technology learning center, including: • Selecting and purchasing hardware and software • Selecting and purchasing instructional materials and other learning resources • Organizing storage and retrieval of children's electronic files • Identifying network upgrade and repair needs • Maintaining a clean and safe physical environment. 4. Assist the director of youth development with grant proposals, site visits with funders, and other fundraising activity as appropriate. 5. Assist in the recruitment and management of other instructional staff, mentors, and volunteers to support teaching and program development activity.

(continued on next page)

Job Description (Director of Technology Learning Programs)

Qualifications	• Excellent leadership, management, and communication skills • Strong commitment to youth and a strong belief in their potential • Teaching experience in a nonprofit or K-12 setting • Experience in designing and implementing creative learning programs • Experience with the application of information technology to educational programs • Experience with the application of computer technologies to research, data storage, data organization, electronic publishing, and communications infrastructure • Ability to develop a team approach to learning and staffing • Experience and competencies in working in a multiracial, multicultural environment
Salary and Benefits	[Fill in salary information.]
How to Apply	[Fill in application information, including an email address and website.]

Overview

The importance of professional development to a successful program cannot be overstated. Finding affordable, effective development opportunities for the staff of youth programs is a creative challenge; but training is like car insurance—you can't afford not to have it.

Untrained, inexperienced staff members and volunteers will struggle and may ultimately fail to deliver good programs. Working with children in an after-school or out-of-school program can be a tremendously isolating experience. Every day staff members and volunteers grapple with hundreds of decisions about how to interact with children, respond to changing needs and situations, and create and run program activities. It is important that they are as well prepared as possible to face these challenges.

Assess Staff Skills: A Starting Point

New staff members and staff members who are new to working with youth and technology should begin by assessing their skill levels in a number of areas. The resulting inventory of skills will help you develop an appropriate individualized professional development program. Often staff members will have a wide variety of skills, and it may be difficult to create a single training program that will fit everyone. Assessment can also allow you to pair up staff members who have mastery in different skills.

Using Worksheet 5.1, rate your staff members (and have them rate themselves) according to the following descriptors:

- Needs to learn—has no level of competency in this particular skill

- Competent—has basic proficiency in this skill

- Can teach other staff members—is capable of transferring this skill to a peer in a one-on-one setting

- Can integrate into lesson—is comfortable enough with this skill and with project or lesson development to be able to weave material into a lesson or activity with youth.

Technical knowledge is only one of the areas of expertise your staff will need, as you will see in the worksheet. Your organization might consider providing training in some or all of the skill areas over the life of your program.

Contents

- Basics of assessing staff skills and creating a staff development program

Audience

- Program directors
- Organizational leaders
- Educational administrators

✓ Reality Check

Project-based learning requires skills and experience—both with the content of the project and with the process of doing a project. For example, if the adult leaders in a program don't have a high comfort level with math, it is unlikely that they will be willing to take on a computer programming project that incorporates advanced math concepts. Personality differences and confidence levels can also factor into adult leaders' decisions about whether to pursue projects. Regardless of skill level, adult leaders must be willing to learn along with youth, and they too must be willing to seek out assistance and guidance. Here are some suggestions that can increase the likelihood that adult leaders will take on and experience success with project-based learning:

☐ Work with activity partners who are specialists in the project content area. If a project can start with a trip to a museum, to a nature center, or to an arts group with expertise in doing hands-on activities with children, that experience will provide background and ideas for the project.

☐ Have adult leaders do projects in a training setting before they do the activities with children. Until they have an opportunity to try out an activity, adults may not realize that they actually enjoy and are capable of leading activities such as editing digital photos, creating web pages, facilitating community-building games, constructing things out of wood, and making hand puppets.

☐ Start with projects that fall within the adult leaders' comfort zones. Projects can be developed around skill areas that fall within adult leaders' prior knowledge and experience. Have adult leaders do a skills inventory; often, personal skills that have tremendous learning potential for youth are overlooked, such as playing a sport, speaking another language, cooking, gardening, working with tools, playing chess, playing a musical instrument, reading maps, or caring for animals. Look for project models and curricular materials that encompass familiar areas—but don't get stuck there. Seek out additional training to ensure that your program is providing youth with diverse experiences and skill-building opportunities.

☐ Learn from the best you can find. If you know someone who has a skill on which you want to base a project, get in touch with that person for suggestions on how to learn more about it. Your contact may be someone found through a website or an email list; a neighbor; a friend; a co-worker; or a specialist at a local business, nonprofit, museum, school, or college.

Worksheet 5.1
Assess Staff Skills

Use this worksheet to create a skill assessment tool for potential and new staff members. Modify the list of skills to include those you think are most important to your program. The skills all relate to starting and running technology-enriched, out-of-school learning programs. No one person can be terrific at everything, but the staff as a whole should have these skills. The list does not include core employee competencies, such as communication skills (e.g., interpersonal skills, speaking, and writing) or basic computer literacy skills (e.g., using a mouse or word processing software).

Name of Staff Person:

Skill Category	Needs to learn	Competent	Can teach staff	Can integrate into lesson
Program definition				
Defining learning and social development outcomes for children				
Creating a program description				
Creating a program schedule: determining how to group children into classes, number of classes, length of class session, frequency of class sessions				
Program management				
Facilitating meetings				
Creating and using organizational reporting systems				
Creating, using, and organizing systems for tracking information				
Using messaging systems: email, voice mail				
Setting professional work expectations				
Creating protocols for responding to emergencies				
Coordinating transportation				
Organizing field trips				
Organizing family events				
Technology management				
Selecting, evaluating, and purchasing software and hardware				
Identifying and selecting a technology support provider				
Troubleshooting desktop and network problems				
Installing and uninstalling software				
Organizing storage of data on a network				
Creating, saving, and deleting files and folders				
Using web browsers (using links and navigation buttons, customizing browser settings, setting up accounts, saving web data)				
Using search engines (typing and copying URLs, searching entire web for information, searching web pages for information, performing custom searches)				
Using web discussion forums (posting, replying to a posting, removing a posting, searching for a posting, using threaded discussion format)				
Using email (setting up accounts, addressing a message, using an address book, making a group address list, composing and editing messages, sending and receiving messages, making and organizing folders, managing inboxes, handling attachments)				

(continued on next page)

Worksheet 5.1 continued

Assess Staff Skills

Skill Category	Needs to learn	Competent	Can teach staff	Can integrate into lesson
Technology management *continued*				
Using electronic mailing lists (finding a list, subscribing to a list, posting a message, replying to messages, unsubscribing from a list, understanding different types of lists)				
Locating and installing plug-ins				
Setting up the physical environment				
Designing display space				
Arranging storage space				
Creating projection space				
Making good use of workspace				
Creating learning stations (e.g., animal/insect table, science table)				
Staff and volunteer recruitment				
Defining full-time, part-time, intern, and volunteer positions				
Creating position announcements				
Identifying recruitment sources				
Interviewing and evaluating candidates				
Developing strategies for recruiting teens				
Creating materials for program recruitment				
Developing and maintaining ongoing program communications (website, electronic news list, and print newsletters)				
Working with children and youth				
Establishing class/program routines and behavior expectations				
Providing positive reinforcement				
Resolving conflicts				
Using effective discipline strategies				
Handling situations of violence and verbal abuse between children				
Instruction				
Modeling (teaching by example)				
Facilitating discussions				
Identifying children's learning styles				
Identifying children with special learning needs				
Presenting information to meet different learning styles (audio, kinesthetic, visual)				
Selecting age-appropriate learning materials (e.g., art supplies, construction or play materials)				
Encouraging cooperative learning				
Setting up partner work				
Setting up group work or work teams				
Instructing on a one-to-one basis				
Providing one-to-one tutoring or homework support				

The YouthLearn Guide • Created by the Morino Institute • *www.youthlearn.org*

(continued on next page)

Assess Staff Skills

Skill Category	Needs to learn	Competent	Can teach staff	Can integrate into lesson
Project development				
Defining a project theme				
Defining a specific inquiry or investigation				
Defining goals for a project				
Defining work products for a project				
Using the Internet to research ideas and resources for projects				
Breaking projects into lessons or class sessions				
Creating a documentation or knowledge capture process for projects				
Selecting software to support project activities				
Lesson development				
Creating a template for lesson plans				
Defining lesson themes				
Defining objectives and skills for lessons				
Designing age-appropriate activities				
Sequencing activities (what to implement and in what order)				
Linking activities (use of themes, transitions, activity extensions)				
Identifying online resources to support activities				
Selecting books and other reading materials to support activities				
Creating a template for class logs				
Using class logs to analyze/improve/extend lessons				
Literacy and communication				
Reading, including setting up a reading group, using read-alouds and silent reading, and teaching technical aspects of reading (i.e., decoding, word recognition, and phonetic understanding)				
Writing (providing lessons in sentence composition, grammar, punctuation, and spelling; using journals; editing; and producing publishable writing)				
Creating drama, improvisations, and oral language exercises				
Cultural competence (understanding differences in languages and cultures)				
Communicating with parents and family members				
Creating opportunities for parent and family participation in program				
Health and safety				
CPR and basic first aid training				
Medical emergency response procedures				
Fire safety				
Making referrals for families in the event of emergency advocacy or assistance needs (legal, housing, food, clothing, medical, and mental health)				
Responding to situations of confirmed or suspected family violence and abuse				

(continued on next page)

Worksheet 5.1 continued

Assess Staff Skills

Skill Category	Needs to learn	Competent	Can teach staff	Can integrate into lesson
Other				

The YouthLearn Guide • Created by the Morino Institute • *www.youthlearn.org*

Staff Development Considerations

Now that you have figured out what kinds of skills your staff needs, how do you go about creating a staff development program? Finding great development opportunities for youth program staff members can be difficult for many reasons:

- **Cost.** Your program may provide few or no resources for training, courses, conferences, retreats, or other forms of staff development.

- **Time.** When training or other development opportunities are available, many organizations find it difficult to release staff members from their day-to-day responsibilities.

- **Shortage.** There is a shortage of quality trainers, models, and materials for both after-school and out-of-school programs.

> " *Training needs can be met in a variety of ways, depending on learning skills, staff requirements, and logistical issues.* "

However, training needs can be met in a variety of ways, depending on learning skills, staff requirements, and logistical issues. Some options for staff development include:

- Providing online tutorials or how-to materials

- Pairing staff with a "tech mentor" for several sessions of on-site training

- Pooling resources with other community organizations to hire a trainer for sessions that will benefit all the organizations

- Exploring professional development sessions that may be offered by the local school district and could be made available to your staff members

- Seeking funding for professional development from local foundations and corporate partners.

Sample 5.1 provides a few examples of staff development sessions to give you an idea of what areas and content they might cover. As you can see, you can choose from a wide range of topics—the only limit is your imagination. However, every session should integrate relevant uses of technology or technology enhancements that might be appropriate for the subject matter.

> @ *For more information, examples, and resources, see the extensive YouthLearn website at www.youthlearn.org*

★ | **Sample 5.1**
Staff Development Sessions

Topic	Content
Fundamentals of child development	• Stages of development for the children in your program, including social, intellectual, psychological, and physical development • How to adapt teaching strategies to children's varying developmental stages • Selection of appropriate learning instruments, instructional material, content topics, and themes • Discussion of situations that challenge healthy development
Inquiry-based learning and the Internet (see Chapter 6)	• How inquiry-based learning differs from traditional teaching approaches • Teaching strategies that support inquiry-based learning • Planning an inquiry-based project and preparing for challenges that arise • Designing projects that focus on the investigation of specific topics and questions developed by the young people in your program • Helping kids learn to ask productive questions • How the Internet can be used to implement inquiry-based learning through email, web searches, and production of web pages
Learning in groups	• Using cooperative learning techniques • Helping kids develop collaboration skills • Teaching kids how to investigate inquiries together • Helping kids work together to map an inquiry and identify a project's goals • Arranging the physical space in your program to encourage learning experiences
How to introduce multimedia tools	• Drawing simple objects using straight and curved lines, angles, dots, circles, and other basic shapes • Using mapping (both on paper and using software) to brainstorm and generate ideas (see Chapter 7) • Introducing photography to the young people in your program, including taking pictures and altering them using software such as Photoshop • Using multimedia projects to foster collaborative learning • Outlining and storyboarding to organize multimedia projects • Introducing HyperStudio to create multimedia presentations
Teaching animation and video, from flipbooks to moving pictures	• How to make flipbooks • How to create simple animation devices • Developing a project through mapping and storyboarding • Producing simple 30-second videos with digital video cameras

The YouthLearn Guide • Created by the Morino Institute • *www.youthlearn.org*

chapter 6

DEVELOP PROJECT-BASED LEARNING STRATEGIES

Overview

The focus of the teaching approach described in this manual is project-based learning, a strategy that combines the concepts of inquiry-based instruction with collaborative learning. These concepts, described in detail in this chapter, are increasingly accepted as best suited to providing children with learning experiences that last—in other words, learning that gives them the knowledge and skills they can use to enhance their lives and achieve success now and as adults. This chapter provides the first part of a road map for creating an engaging, project-based curriculum for your technology learning center.

Understand Project-Based Learning

In project-based learning, children conduct investigations on topics of relevance and interest to them. The projects last several weeks and involve myriad activities, all related to a central theme or topic of inquiry. Students usually work in groups to conduct the investigation and then to communicate what they have found out in some tangible way, such as a poster, multimedia presentation, book, or other product or production. In a school setting, in which specific curricular goals must be met, a teacher can establish the topic or general theme for projects. In out-of-school learning programs, children can have more freedom to identify their own questions. In either setting, children should have a role in framing their questions and in determining how to go about finding the answers. The teacher's role is a supporting one—as a guide and facilitator of the process rather than a director and provider of answers.

Contents

- Tips for adopting a project-based approach to teaching
- Advice on how to create relevant and interesting projects
- Tools for creating project-based lesson plans

Audience

- Youth development staff
- Teachers and center instructors
- Program directors

Snapshot

Project-based learning and outdoor education are standard components of the curriculum at the Eagle Rock School (www.eaglerockschool.org), a tuition-free, year-round residential school for teens who have not succeeded in traditional school environments. Projects are interdisciplinary, incorporating math, language arts, social studies, and science. Students might build a cabin in the woods while reading Henry David Thoreau's *Walden*, do a presentation on the community organization strategies behind the AIDS Quilt, or create a proposal for a project to implement at a local nonprofit after researching the nonprofit's history, mission, and operating structure.

What Is Inquiry-Based Learning?

The inquiry-based approach is rooted in the belief that people learn best when they construct knowledge for themselves, building on experience, current knowledge, and an active process of uncovering meaning through investigation. The process therefore starts with the learner's own questions. The emphasis is not on teaching facts and concepts but on giving learners the tools they need to build their own knowledge.

Inquiry-based learning projects are not *unstructured*, however—they are *differently* structured. In fact, inquiry-based learning may require even more planning, preparation, and responsiveness on the part of the instructor than does traditional instruction. In an inquiry-based learning approach, instructors help children identify and refine their "real" questions into learning projects or opportunities, then guide the subsequent research, inquiry, and reporting processes. Because one role of out-of-school programs is to enhance, support, and expand on the core curriculum of K-12 schools, inquiry-based learning is a particularly good approach for giving kids an opportunity to learn with more freedom while reinforcing basic skills they are learning in school.

Inquiry-based learning has other advantages as well:

- **An inquiry-based approach is flexible.** It works well for projects that range from the extensive to the bounded, from the research-oriented to the creative, from the laboratory to the Internet. (It is essential, however, that you plan ahead so that you can guide the kids to suitable learning opportunities.)

- **It works for children with different learning styles.** You'll find that many kids who do not respond well to lectures and memorization blossom in an inquiry-based learning setting, gaining confidence, curiosity, and self-esteem.

- **It encourages interdisciplinary thinking.** The traditional approach tends to be highly segmented. A class might study science for a while, then language arts, then math, then geography. In contrast, the inquiry-based approach is at its best when being used for interdisciplinary projects that reinforce multiple skills or knowledge areas in different facets of the same project. You'll also find that although the traditional teaching approach is sharply weighted toward the cognitive domain of growth, inquiry-based learning projects reinforce skills in cognitive as well as physical and social-emotional domains.

- **It is well suited to collaborative learning environments.** You can create activities in which the entire class works on a single question as a group (just be sure that the whole group truly cares about the question) or in teams working on the same or different questions. Of course, it also works well when you've decided to let each student develop an individual project; if you do, however, be sure to incorporate some elements of regular collaboration or sharing.

- **It can work with any age group.** Even though older students will be able to pursue much more sophisticated projects, instructors should build a spirit of inquiry into activities with even the youngest children.

What Is Collaborative Learning?

The term "collaborative learning" refers to a teaching strategy in which students work together in small groups toward a common goal (usually a product representing all the students' contributions). The students are responsible for one another's learning as well as their own. Thus, the success of one student helps other students succeed.

The concept of collaborative learning has been widely researched and advocated throughout the education field. Research on collaborative learning has shown that the active exchange of ideas within small groups not only increases interest among the participants but also promotes critical thinking. There is persuasive evidence that cooperative teams achieve at higher levels of thought and retain information longer than children who only work quietly as individuals.

The notion of collaborative learning can extend far beyond children and instructors; it can also mean involving parents, the local community, and even people across the globe in the learning process. Advances in technology have provided powerful new tools and techniques to use in communicating and collaborating far beyond a traditional classroom. There is no reason, for example, why your students cannot use the Internet to collaborate with children in another part of the world, one that they are interested in learning about.

Develop an Inquiry-Based Project

The essence of inquiry-based learning is that children participate in the planning, development, and evaluation of projects and activities. Inquiry-based projects must fit into your larger program goals, of course, but students should be actively involved in planning the projects and asking the questions that launch their inquiries. Developing most projects involves three basic steps:

1. **Pre-planning.** Before working with the kids, determine any preliminary factors or characteristics that must be in place for you to achieve your larger goals or plans. Consider factors such as scope, the amount of time you'll spend over how many sessions, relationships to other projects, topical focus, age appropriateness, skills you want to use, resources, media, and collaboration techniques. Make any decisions up front that you have to, but remember to let the children participate in the planning process as much as possible.

2. **Brainstorming.** Assuming the widest range of possibilities, start a discussion in class to find out what the kids are interested in. Ask some broad questions about their interests. Try some simple brainstorming activities to record the ideas they suggest and to begin paring them down to one or a few (see Chapter 7).

3. **Questioning.** Almost any topic can become the foundation for an inquiry-based project, even something as mundane as shoes, if that's what the kids are interested in. Suppose you've decided on that topic. Ask the kids what they would like to know about shoes, and map the questions to areas of study as shown in the curriculum wheel illustrated in Figure 6.1.

A curriculum wheel on the topic of shoes

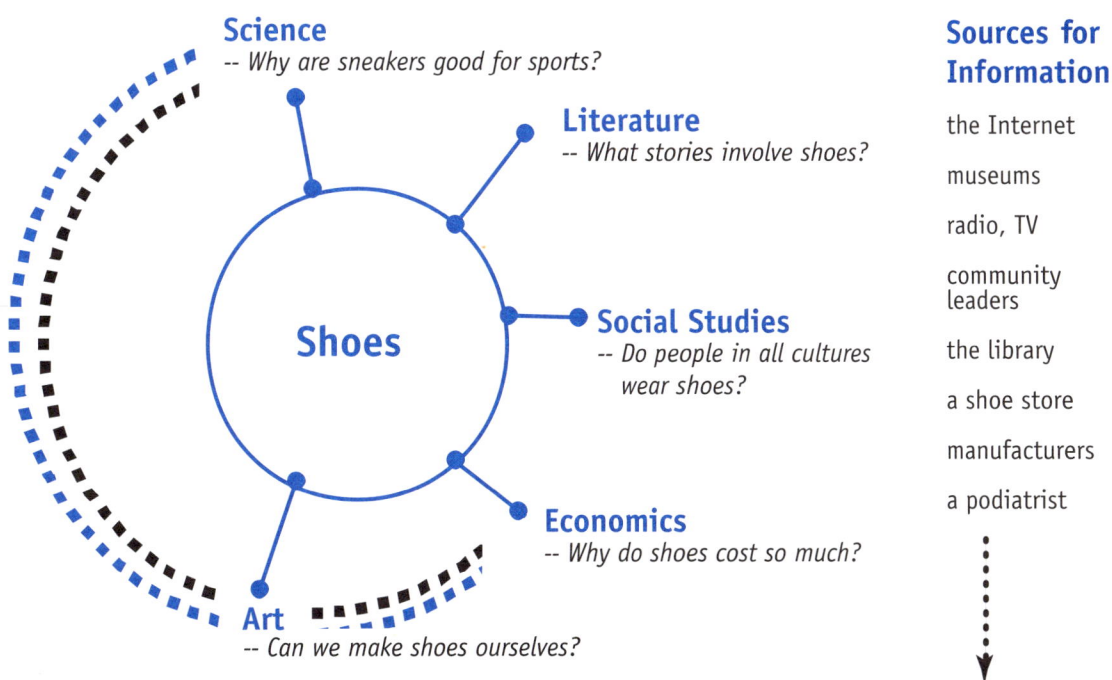

Science
-- Why are sneakers good for sports?

Literature
-- What stories involve shoes?

Shoes

Social Studies
-- Do people in all cultures wear shoes?

Economics
-- Why do shoes cost so much?

Art
-- Can we make shoes ourselves?

Sources for Information

the Internet

museums

radio, TV

community leaders

the library

a shoe store

manufacturers

a podiatrist

Figure 6.1

As you work with your students to develop a project, ask questions like "Where could you find resources to answer your questions?" Incorporate visual mapping techniques to select and refine questions and associated projects or activities. Remember, your job is to guide the kids as they navigate the learning process for themselves.

In choosing class projects, avoid letting individuals work alone on totally unconnected projects. It's not that there's anything wrong with working alone, but the kids won't get the advantage of developing collaboration skills, and you'll be spread awfully thin trying to help them all on such disjointed topics. In most cases, you'll be better off having the whole class work on a single concept or breaking up into teams to work on particular questions, aspects, or executions of that theme or idea.

Teachers can take many approaches to crafting an inquiry-based project, but there are four essential parts: posing real questions, finding relevant resources, interpreting information, and reporting findings. The process is illustrated in Figure 6.2.

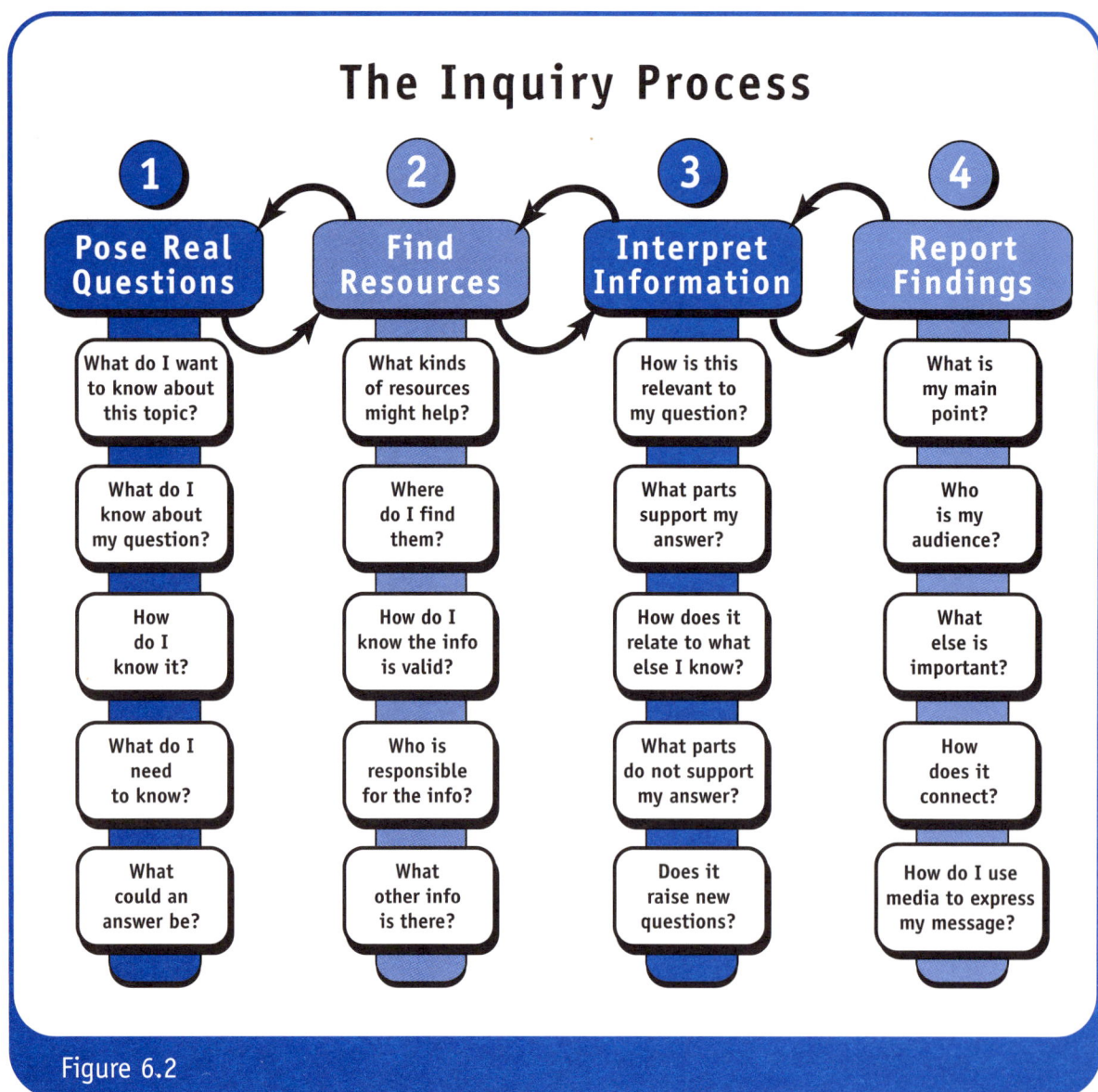

The Inquiry Process

1 Pose Real Questions	2 Find Resources	3 Interpret Information	4 Report Findings
What do I want to know about this topic?	What kinds of resources might help?	How is this relevant to my question?	What is my main point?
What do I know about my question?	Where do I find them?	What parts support my answer?	Who is my audience?
How do I know it?	How do I know the info is valid?	How does it relate to what else I know?	What else is important?
What do I need to know?	Who is responsible for the info?	What parts do not support my answer?	How does it connect?
What could an answer be?	What other info is there?	Does it raise new questions?	How do I use media to express my message?

Figure 6.2

Step 1: Pose Real Questions

Helping students arrive at their "real" questions is the central technique of inquiry-based learning. It involves encouraging kids to ask themselves:

- What do I want to know about this topic?

- What do I need to know?

- What do I know already and how do I know it?

- What might a possible answer be?

The goal is not to ask just any questions, of course, but ones that children honestly care about. Because inquiry-based learning is premised on helping children ask questions, learning the art of asking good questions is a primary skill. Good questions share the following characteristics:

- **The question must be answerable.** "What is the poem 'Dream Deferred' based on?" is answerable. "Why did Langston Hughes write it?" may be answerable if such information exists or if the students have some relevant and defensible opinions. "Why did he choose this particular word in line six?" is not answerable because the only person likely to know such a specific answer is Hughes himself, now deceased.

- **The answer should not be a simple fact.** "In what year was Lincoln killed?" doesn't make for a very compelling project because you can just look up the answer in any number of books or websites. "What factors caused the assassination attempt?" might be a good project because it will require research, interpretation, and analysis.

- **The answer should not already be known.** "What is hip-hop music?" is a bit too straightforward, and the kids are not likely to learn much more than they know already. "What musical styles does hip-hop draw from and how?" offers more opportunity for exploration.

- **There must be some objective basis for an answer.** "Why is the sky blue?" can be answered through research. "Why did God make the sky blue?" cannot because it is a faith-based question. Both are meaningful,

valid, real questions, but the latter isn't appropriate for an inquiry-based project. "What have people said about why God made the sky blue?" might be appropriate. Likewise, "Why did the dinosaurs become extinct?" is ultimately unanswerable in that form because no one knows for sure, but "What do scientists believe was the reason for their extinction?" or "What does the evidence suggest about the cause?" will work. Questions based on value judgments don't work for similar reasons. You can't objectively answer "Is *Hamlet* a better play than *Macbeth*?"

- **The question must not be too personal.** "Why do I love the poetry of W. B. Yeats?" might inspire some level of internal exploration, but in most cases that's not your most important goal. Get the kids to focus on external research instead.

When working with younger, shy, or alienated kids and with those unused to this sort of approach, you may have to ask leading questions or even spoon-feed them questions to get started. Don't be discouraged. Once they catch on, you'll see their enthusiasm and curiosity grow.

✔ Reality Check

Children with special emotional needs may have an enormously difficult time handling the responsibilities of collaborative learning. If you are working with a child who exhibits clear signs of abuse or extreme stress—a child who presents a physical threat to himself or herself or to others through repeated hitting, pushing, or throwing; a child who repeatedly uses violent or sexually threatening language with other children; a child who repeatedly demands an unreasonable amount of individual attention from adults; or a child who is severely socially withdrawn from other children—get help. Consult an experienced social worker or mental health professional on how to interact with the child and how to make appropriate referrals for intervention, if needed. Get advice from an experienced teacher or staff person on how to structure program activities to support and include the child appropriately.

Make sure the inquiry relates to a student's real question; be careful to avoid "bait and switch" ploys in which the student actually pursues the teacher's interests. Helping students find their real question can sometimes be no easy matter because they may not know the real question themselves. For example, a student might tell you that he is interested in how streets are named or who names them, but what he might really be interested in is learning if he could ever have a street named after him. Through skillful questioning, you must help students find out what they really want to learn so that you can focus the inquiry. This is particularly important with young students, who do not have fully developed reasoning skills.

You must then help students identify and refine their questions for exploration and help them realize when a question is not appropriate for a given project. The process of refining questions includes helping students identify what they know and don't know about the subject, identifying subquestions that may be part of the larger question, and, most important, formulating hypotheses about what the answer might be. This last step can be a powerful tool in determining whether a question is answerable.

Step 2: Find Relevant Resources

Between the question and the answer are sources of information. What kinds of sources might help? Where do you find them? How

Tip

Asking Good Questions

Be aware of how a question can either shut down or open up a conversation by the words you choose and the prejudices you reveal. For example, consider the different responses you'd get to the question "Nobody here has ever created a web page, have they?" versus "Has anyone made a web page before?" versus "What do we know about creating web pages?" The second question is at least a more positive version than the first, but it still will only get you yes or no answers. The third invites constructive input and validates prior knowledge.

do you know that the information is valid, and who is responsible for it? What other information is there? Answering such questions begins the process of assembling and then assessing evidence to find an ultimate answer to the inquiry. The key distinction in this phase is that the focus should be not on finding the answer but on finding sources that might have information that could lead to the answer.

Sources of information include books, people, experiments, web pages, and discussion groups on the Internet. Finding source material sometimes is the easiest part of the process; assessing the information is trickier. Because real inquiries are rarely about objective facts, it is probable that no one has ever posed the inquiry in exactly the same way before. Finding a swift, pointed answer is not likely to occur.

Children must be taught how to collect bits of partial answers and assess their validity. Because all information tends to be biased by the perspective, experience, or interest of its author—whether it's from a book or one's grandmother—developing critical evaluation skills is key. Not everything that comes from an established publisher is good information, and not everything that comes from a personal home page is bad information. The kinds of things that kids ask may be answerable only

Snapshot

Youth participating in after-school and summer programs at the Eli Whitney Museum (www.eliwhitney.org) make crafts and simple machines ranging from wooden boats to hand puppets to model airplanes. Projects often start with a question or design challenge, such as how to design a wooden device that will carry a marble from point A to point B. A "crash test dummies" class explores how to design a battery-powered car, wooden kayak, and parachute that will safely hold a dummy. Courses in programming games and developing websites are offered in a computer lab.

by other people, perhaps only by other kids. Perhaps all the published information is worth relatively little because it was written by adults and does not address the specific questions of young people.

Evaluating sources of information is especially important in light of the widespread misinformation in our world, misinformation made all the more accessible by the Internet. An enormous amount of valuable and exciting information is available on the Internet, but an enormous amount of total nonsense, falsities, half-truths, and unsupported theories is also out there. Kids have to learn to distinguish between the two, but you can't give them hard-and-fast rules. Be careful, especially with younger children, to instill a healthy awareness of the concept that information is "authored" rather than a broad distrust or even disdain for authorities.

Step 3: Interpret Information

Closely related to evaluating the quality of information is the next step: evaluating its applicability to the original question. How is this information relevant? How does it relate to what else we know? What parts support the hypothesis, and what parts do not? Again, students catalog information and record new questions that arise, but in this step, they focus on the relationship of that information to the hypothesis and to the other bits of information they've collected.

One critical aspect of the interpretation process is convincing children of the need to look actively for information that disproves their hypothesis. Just because some data support the hypothesis does not mean it is correct. Other explanations could apply to the same data.

Don't confine yourself or your students to the web when doing research on the Internet. The Internet's most powerful feature is the ability it gives you to reach real people through email, chat, discussion boards, and online events. Do the kids have a biology question? Make sure they do the necessary background research first, but if they still have questions, try visiting the website of a local (or even national) college or university. You'll probably find email addresses for faculty members, or at least an address for

💡 Tip

Templates Can Help With Developing Inquiry-Based Projects

Cornelia Brunner of EDC's Center for Children and Technology has developed a series of simple planning templates to help teachers develop inquiry-based projects with their students:

- Getting started (identifying preliminary questions and information)

- Planning (specifying unknowns and hypotheses)

- Focusing (refining the primary question)

- Identifying and mapping possible resources

- Evaluating resources

- Tracking what fits and what doesn't

- Assessing preliminary answers

- Making sense of multiple pieces of information.

These templates can be found at www.youthlearn.org.

each department. Have the kids go right to the source for help, and they'll probably get it as long as they are polite and not too demanding.

During the validation and interpretation processes, students continue to refine their real question and, one hopes, find an answer to it, although that answer may not be as simple as they originally thought. It may even contradict their original hypothesis, but that should not be a discouraging experience if it is handled properly. Think about video games: Instead of being discouraged when they hit an obstacle, kids see the obstacle as a challenge to overcome and will go to great lengths to do so. Instructors should emphasize at each stage of the inquiry that the investigation is a journey and that finding new information is exciting, especially when unexpected.

Step 4: Report Findings

Even the reporting stage in an inquiry activity contrasts with traditional educational methods. The emphasis should be on telling a particular audience the personal story of the "learning journey," rather than just recounting the facts as in a traditional paper. The objective is not to state the answer but to tell how the students arrived at this answer.

With highly cognitive projects, don't be overly concerned if kids have gaps in their basic skills, such as reading or spelling. Yes, it will cause obstacles and complications, and you must be sure not to set the kids up for failure by pursuing projects that are beyond their age or abilities. However, one of the great advantages of the inquiry-based approach is that kids will want to learn the answers and will become more energized to overcome any skill deficits to achieve that goal. Don't derail their progress at this point by fretting over every misspelled word or grammatical error.

The finished product(s) for the project could appear in any number of forms—a paper, a web page, a collage, or a slide show, just for starters. When is it time to report? Since students are dealing with self-directed questions that have highly personal value, they should report when they are satisfied with the answer.

Create Lesson Plans

To organize your project into individual sessions, you should create detailed lesson plans. The process of preparing them helps instructors organize their thoughts for each day's work with children. They provide documentation that becomes the basis for reflection and future refinement of the instruction process. They also enable instructors to capture specific teaching strategies in a format that is easy for others to understand and follow. If multiple instructors, volunteers, and interns are working with children in a single class, creating a lesson plan ensures that everyone knows how and when and why the activities are being done.

In project-based instruction, many lesson plans would be developed over the course of a multiweek project. In an after-school or other out-of-school context, for example, a lesson plan would describe what happens

Snapshot

Looking for live inspiration on inquiry-based learning? If your area has a children's museum, take a visit—with a notebook in hand. The Boston Children's Museum (www.bostonkids.org/exhibits/index.htm) is one of the best of its kind. Exhibits range from replicas of a house in Japan to bubble blowing to a pretend grocery store. If the museum in your area offers teacher curriculum guides or activity books for home, pick up a few copies. A science museum with hands-on exhibits for children will offer similar inspiration.

with one group of children in one session. A project, in contrast, is a series of interrelated lessons implemented over several sessions.

A lesson plan is a working document; it should be revised and updated as necessary to reflect changes made once the plan has been used. Your learning center may want to assemble an ongoing "best of" collection of lesson plans that have been rewritten to reflect how they were actually implemented or should have been implemented.

Because a lesson plan is first and foremost a personal planning tool for an instructor, each instructor should use a format that works best for him or her. At a minimum, however, lesson plans should include the following details:

- Age of children
- Activity steps and procedures
- Strategy for incorporating the Internet and related technologies
- Goals for the session (what children will accomplish or produce by the end of the session)
- Materials to be used
- Preparatory steps (things to be done before the children arrive).

Lesson plans might also include the following components:

- Introductory activities

- Transitional activities (activities that bridge a change of activity or a physical move to another space)

- Closing activities (activities that help children process what they have learned and prepare them for the next day's work).

You don't have to create your own lesson plans from scratch. Countless websites allow you to download plans that others have created and used. You can find good information in unusual places. For example, a lesson plan posted by an individual—say, a hobbyist in a particular subject—might be wonderfully suited to a project you are designing. Conversely, you may find that a lesson plan posted by a recognized source, such as a university or museum, is not appropriate for your needs.

When you *do* create lesson plans from scratch, you will find the Internet to be a wonderful place to find relevant source material for your students. But whenever you use the Internet for this purpose, keep the following considerations in mind:

- **Follow the links.** Before using a site in a lesson, carefully read through all the pages, test the links to make sure they are active, and investigate the links and the source of the site to ensure that children are not exposed to inappropriate material.

- **Make your own page of links.** Create a simple web page of links to sites that will be used in your lessons. Doing so will decrease the amount of class time spent looking for sites and reduce the risk that children will encounter inappropriate material. This kind of page can become the starting point for a series of web pages that children build themselves.

- **Consider searching as an art in itself.** Searching for information is an important skill that involves analysis, associative thinking, and competency with search engines. Especially with younger children, instructors may want to introduce students very selectively to specific sites and allow

kids to become comfortable with basic web browser functions before teaching them how to search on their own.

- **Look for content made by children.** Child-generated content is in abundance on the Internet. Whenever possible, direct your students to sites that feature children's original drawings, photographs, writing, video clips, or audio clips. Such sites provide a model for your students' own "publishing" and generate ideas for themes, styles, and techniques.

Sample 6.1 will give you a better appreciation for what a successful lesson plan looks like (for more lesson plans, see Chapter 9). After reading some sample lesson plans, you can use Worksheet 6.1 to create your own lesson plan. Worksheet 6.2 provides a useful checklist of creative materials you may want to have on hand when you begin to do projects with your students.

@ *For more information, examples, and resources, see the extensive YouthLearn website at www.youthlearn.org*

Develop Project-Based Learning Strategies

⭐

Lesson Plan (for 5- to 8-year-olds)

Lesson Title: Camera Investigations	Age Group: 5- to 8-year-olds

Goals for the Session	• Learn about animation and visual storytelling through production of a zoetrope (a simple animation tool) • Practice cooperative interaction • Practice reading and writing • Practice mouse and web browser skills • Practice listening, taking turns, sharing • Practice drawing • Practice planning—predicting an action and representing it visually
Materials	• *Like Likes Like*, by Chris Raschka (children's book) • Large index cards • Pencils • Easel paper • Fine-tip colored markers • Manila folders • Scissors • Plastic lazy susan (for zoetrope) • Cardboard strip, enough to create zoetrope at least 10" in diameter • QuickTime software

Preparation	• Prepare web page with links to these sites:
	
	   • *How to Make a Zoetrope*, by Ruth Hayes, from the Random Motion website (www.randommotion.com/html/zoe2.html)
	
	   • Videos by students at Hoffer Elementary School in Banning, California, from the California Museum of Photography website (www.cmp1.ucr.edu/exhibitions/hoffer/home/hoffer.video.html)
	
	   • Interview with Ashley T., a project of Room 36 at Lincoln Bassett Elementary School, from the LEAP website (www.leap.yale.edu/lclc/town/class/ashleyt/ashleyt.html)
	
	• Place web page shortcut icon on desktops.
	
	• Download QuickTime software onto at least five workstations.
	
	• Charge up digital camera batteries.
	
	• Make zoetrope.
	
	• Cut paper to make zoetrope animation strips.

(continued on next page)

Lesson Plan (for 5- to 8-year-olds)

Lesson Title: Camera Investigations   **Age Group: 5- to 8-year-olds**

Lesson Steps

1. Introductory activity (5 min.)

Fruit Basket: The group sits on chairs in a circle, with one person standing in the middle. Participants are assigned to one of three groups named for fruits (such as apples, oranges, and pineapples). When the middle person calls out a fruit (e.g., "apples"), all the apples change chairs, including the middle person. The person "out" becomes the next caller. If a caller says "fruit basket," all group members have to change chairs.

2. Word for the day/reading and writing exercise (20 min.)

While sitting in a circle, the children each have a turn to give one word that is "their" word. It could be a word that describes a feeling, something that is special or important to the child, or something that he or she is thinking about. Write each word on an index card as it spoken and then give the card, a pencil, and a manila folder to the child. Children work in groups of two or three to write a sentence about their word on the other side of the index card. Work with each group to help pre-readers spell the words for their sentences and to facilitate group interaction. Groups should be balanced to include at least one independent reader/writer in each group. Each child should store his or her card in the manila folder.

3. Read-aloud (10 min.)

Read *Like Likes Like,* by Chris Raschka. (This book was selected because it has great illustrations; the story is told with a few simple words; it evokes several themes, including loneliness, companionship, difference, and sameness; and it is an excellent example of visual storytelling.)

4. Transitional activity (5 min.)

Frozen-in-Motion: Sit in chairs with the students. Initially, have them feel the floor, feel the chair, and feel the space they are in. This can be done with eyes open or closed. Then tell the students to feel and replicate an emotion, such as boredom, surprise, or anger. The leader (you or a student) then says, "Freeze!" Everyone freezes like statues, and the leader now says, "We are in the museum of boredom" (or surprise or anger, whatever the selected emotion is). The leader selects one person to hold his or her frozen pose and be the "statue." Everyone else focuses attention on that person and describes different aspects of the statue, such as the posture or facial set. This is a great exercise for training observation skills and is good for building enhanced vocabulary for writing.

5. Review videos made by children (15-20 min.)

Show the group how to pull up a video web page on a computer and demonstrate how to download and manipulate video clips. Note how each clip consists of a series of frames—still images—that, when run together, create the motion we see. The group divides into pairs to look at video clips on Hoffer and Room 36 websites.

6. Make zoetrope animations (30 min.)

The group views the model zoetrope and sample zoetrope animation strips made by an instructor. Note how the animation was created in frames—think of a simple motion, draw the first movement, then draw the final movement, and fill in the gradual changes in between. Each child then receives a strip of paper to make his or her own zoetrope animation with the colored markers. The group sits together on the floor and shares markers. Paper and scissors should be available for children to make additional animation strips if they need them.

7. Closing activity (5 min.)

Spend a few minutes wrapping up the session by talking about it with students. Ask open-ended questions to help them assess what they have learned, and be sure to say what you have learned, too. For example, say, "Today I learned how to use the zoom on the digital camera"; then ask the group to share what they have learned, what they liked or didn't like about the session, what they found out about a fellow group member, and so forth.

Worksheet 6.1
Create Lesson Plans

Read the previous section and lesson plan carefully, and then use this template to create your own lesson plans. Using templates like this one will help ensure a high level of consistency and structure in your program.

Name of Lesson	
Instructor(s)	
Age of Children	

Goals for the session (What will children accomplish or produce by the end of the session? What skills and competencies will children practice or develop?)

Materials needed

Preparation (List the activities and tasks that must be done prior to the class session.)

Activity steps

Amount of time:	**Introductory activity**
Amount of time:	**Step 1** (Include strategies for incorporating the use of the Internet and related technologies in each step, as appropriate.)
Amount of time:	**Step 2**

(continued on next page)

The YouthLearn Guide • Created by the Morino Institute • www.youthlearn.org

6

Activity steps *continued*

Amount of time:	**Step 3** (Add additional blocks for other steps, as needed.)
Amount of time:	**Transitional activities** (Add activities that bridge a change of activity or a physical move to another space; insert in lesson plan as needed.)
Amount of time:	**Closing activity** (Add an activity that helps children process what they have learned and prepares them for the next day's work.)

Assessment (What indicators will determine what children have learned from the lesson?)

Resources used (List the books, reference materials, websites, etc., used in creating and carrying out the lesson plan.)

Worksheet 6.2
Checklist of Creative Materials

A technology learning center is first and foremost a classroom. Therefore, you will need many types of supplies that have little to do with computers. You may be limited by practical concerns like space and budgets, but try to keep a supply of the items on this checklist for near-everyday use.

Essential Items

- [ ] Large white pads of paper (at least 2' x 3') that can stand on easels and on which you can write during mapping and webbing activities and other demonstrations. (These pads are better than chalkboards because you can detach the used sheets and tape them to the wall.)
- [ ] Traditional composition books with which to make journals
- [ ] Fine-point felt-tip marker sets
- [ ] Thick-point felt-tip marker sets
- [ ] Crayons in boxes of 24
- [ ] Colored pencil sets
- [ ] Colored chalk in boxes or tubs
- [ ] Index cards in packs of 100
- [ ] Posterboard in packs of 10 sheets
- [ ] Butcher paper rolls
- [ ] Construction paper in multicolor packs
- [ ] Glue sticks
- [ ] Post-it notes in multicolor packs
- [ ] Scissors, child size
- [ ] Scissors, adult size
- [ ] Rulers, 12", plastic
- [ ] Hole punches
- [ ] Staplers and staples
- [ ] Color folders in packs of 100
- [ ] Pencils
- [ ] Pencil sharpeners
- [ ] Rolls of transparent tape
- [ ] Rolls of masking tape

Great Additions

- [ ] Plastic storage bins
- [ ] Gel pens
- [ ] Modeling clay and tool pack
- [ ] Binding machine (for making books)
- [ ] Plastic bindings
- [ ] Clipboards
- [ ] Tracing paper
- [ ] Yarn
- [ ] Finger paints
- [ ] AA batteries (for cameras, calculators, etc.)
- [ ] Poster paints
- [ ] Paintbrushes
- [ ] Glitter and other adornments
- [ ] Map of the world
- [ ] Map of your country
- [ ] Corkboard

chapter 7

DEVELOP EFFECTIVE
TEACHING TECHNIQUES

Overview

Regardless of what topic your center's staff members are teaching or the age of the kids in their classes, certain effective teaching techniques should become a routine part of instruction in your program. These techniques, which range from good modeling to the daily use of journals, can help you inspire in your students a love of learning, a sense of wonder, and a habit of reading. They will also help you create a stable, structured environment where kids feel comfortable exploring new ideas and learning new skills.

Techniques for Good Modeling

Perhaps the most important technique for instructors to understand and master is the art of modeling. To a degree, modeling is simply a fancy word for a very simple concept: People, especially kids, learn more from what they see than from what we tell them.

Modeling is about demonstrating everything you want kids to do in the way you want them to do it. Sometimes modeling involves a specific task, like showing kids how to hold a camera, but it also involves reinforcing global values (e.g., teamwork) through the words you choose and how often you involve others in your demonstrations.

Instilling Values and Attitudes

Kids are remarkably intuitive. From your actions, emotions, and attitude, they know when you truly value something and when you don't. You can talk all day long about the importance of sharing ideas, but if they never hear you ask for input, they'll know you don't really mean it.

That's why anything you consider an important part of your learning programs must be reinforced through everyday activities and throughout your center's environment. All instructors want to inspire a love of reading in kids, for example, so do you read aloud every day? How about quiet time when kids can read their own books themselves? During those "free reading" times, the kids must see you reading your own book or magazine, not working on lesson plans or preparing for a future project. Even though those are important things, kids will get the idea that reading is nothing but busywork to keep them occupied while you're doing more important things. If reading is to be seen as fun, valuable, and important, you must model it that way—every day. Repetition and practice are the keys to reinforcing your model.

Contents
- Advice on how to teach through actions, not just words
- Ideas for encouraging the sharing of ideas, keeping kids focused, and integrating reading into daily activities
- Advice on using graphic organizers and journals
- Tips on teaching about (and with) technology

Audience
- Youth development staff
- Teachers and center instructors
- Program directors

You are always modeling, whether you intend to be or not. Suppose you are doing an image-editing project in Photoshop. If you are tentative because you're insecure about your own knowledge, kids will feel that; you will be modeling fear and anxiety. However, if you show a positive attitude of exploration and learning together, you will model confidence, curiosity, and collaborative learning, even though you don't have all the answers. "So what if we make a few mistakes!" should be the approach. Every obstacle can be a learning opportunity.

When you hand a child a digital camera and say things like "Now be careful, this is very expensive," you show that you are insecure about his or her ability to handle it. If instead you show the kids how to hold the camera properly and confidently—and use that same method yourself every single time—you've modeled trust.

Providing Step-by-Step Demonstrations

Whether you're doing a journal exercise, a lesson on how to conduct an interview, or a web page activity, demonstrating how to do the task will be the most effective way to convey the information to the kids. You must be sure to engage the class and have a way of checking that they really understand what you show them. For most activities, use a pair-share process as much as possible. (See "Teaching the Pair-Share Technique," page 91.) It will take a little longer, but doing so builds community, models good interaction skills, and gives you a way to see whether the kids have really grasped the concept.

Whenever you're demonstrating, keep up the verbal patter, especially by asking lots of questions, even if they're rhetorical: Did you see that? Why do you think I did it that way? Have you ever seen that before? Isn't that neat? This is especially important when you are doing technology demonstrations, such as showing a piece of software, when kids have to sit still and watch for a while. Always keep demonstrations bounded to just a few concepts so as not to overwhelm the kids, and involve them as much as possible.

Remember that everything you do in front of the class has two levels. The first level is the specific activity you're engaged in, and the second consists of the social, cognitive, and community skills you're modeling.

Giving Good Directions

Although modeling is one of the most important techniques for working with kids, it has a potential negative effect that you must avoid—conformity. Kids often feel that they have to please adults, and if you're too strict in your models and directions, you run the risk of blunting their creativity and spirit of exploration. That's why it's important for you to focus simply on introducing concepts and, especially when teaching technology topics, to give kids just enough guidance for them to feel confident while leaving plenty of time for individual discovery.

When you model carefully, kids will tend to copy what you do, so you must decide which tasks are important for them to do a certain way and which they have freedom to experiment with. Your modeling and directions should reflect this difference. Here are some tips for giving directions:

- **Use precise words when you want precise responses or actions.** Telling someone to "hand" the camera to you has a more clearly defined meaning than "give" or "pass," especially if you've consistently used the word "hand" when demonstrating the proper procedure.

- **Be intentionally unclear when there's a larger lesson involved.** For example, if you ask kids to write the names of people in their community, don't tell them what a community is, even if they ask. Part of the activity may be to learn what we each think community is. The key is knowing your goal.

- **Walk around the room offering help and guidance.** This is especially important when you have kids working on an activity that offers them creative leeway. Hold up examples of work in progress so others can see. Perhaps you've asked the kids to "show how we hear things" in support of an audio project you're about to undertake. One child draws a picture of himself with oversized ears. Show it to everyone and say, "We hear with our ears." Another student writes a list of phrases like "from TV" or "friends tell us." Make sure everyone sees that these responses are correct, too. Show the most literal interpretation first, because that's what most kids will have

done. You're trying to open doors of creativity, not bust them down, so your modeling should make the kids feel secure in what they're doing.

- **Verify that you have given clear instructions.** As you walk around the room, you may see several or even many children doing things in a way that is simply incorrect. This strongly suggests that your directions were not clear enough.

⚡ Tip

Teaching the Pair-Share Technique

Step 1: Demonstration Pair.

Select a volunteer to come up front. Show the class what you did (e.g., if you've just done a journal activity to open the day, explain what you did, holding up your book so everyone can see). Appear to be talking mostly to your partner, although you're really talking to the whole class. Then ask your partner to share her work—she'll probably do it much as you've modeled it.

Actively interact and share with your partner to model how people get ideas from each other. For example, in recapping a journal activity, say something like, "Wow, that's a good idea! Can I borrow it?" Then add it to your own work by writing or drawing it in your journal, and let everybody see you do it.

When you're demonstrating new skills or techniques, use a similar process. If you're showing kids a new piece of software, for example, select a partner to repeat your work in front of the group, in his or her own file, after your demonstration. In addition to its other benefits, demonstrating in this way will help make sure that you don't introduce too many topics at once. Always ask lots of questions during your demonstration, and give good directions.

Note: When you're doing a pair-share activity, always start by modeling the deciding of who goes first. Ask the class before you begin sharing your work, "How can we decide who goes first?" They'll say things like "ask who wants to" or "flip a coin" or "ladies first." Take one of their suggestions. It teaches them respect and communication skills.

Step 2: Model Pair.

Next, ask two other students to share their work with each other aloud in front of the class. Note that although they are sharing in front of the group, encourage them in your instructions to share with each other, even though they're doing it aloud. Listen to them as they explain their work to see if they understood your instructions and the concepts involved.

Step 3: Class Pairs.

Finally, have everyone turn to someone else and share his or her work. Give them one or two minutes, then walk around, listening and participating. Keep alert for students who seem to be having difficulty so that you can help them later. If someone has done something exceptionally creative, hold it up for the group. If the students seems to be confused, re-model the activity or technique before moving ahead and repeat the entire pair-share process, paying particular attention to your directions and areas in which people seemed to have difficulty.

Get everyone back on track as a group, and don't single anyone out with individual negative feedback.

- **Be a bit more restrictive early in a project, then more open.** When introducing photography over the course of several days, for example, you may want to restrict the kids at first to taking a certain kind of picture (say, close-ups or angled shots), but then give them the freedom to use a wider range of techniques for the final project.

- **Set limits occasionally.** If you have been encouraging experimentation in all things, kids will not feel restricted when you give them specific directions. In many ways, modeling is all about setting expectations. If over the course of the session you have been modeling experimentation, curiosity, and creativity, kids will hardly notice occasional limits.

- **Keep your words consistent with your actions.** If you tell kids to draw pictures or write words for a mapping activity but all you ever do is write words, they'll get the message and will follow what you do.

Techniques for Encouraging the Sharing of Ideas

In today's networked world, teamwork is more than just a laudable goal—it's a required skill. If you're going to create a truly collaborative environment, you have to model, instill, and reinforce the sharing of ideas in all aspects of your interactions with children. The following is a list of techniques for infusing a spirit of sharing throughout all activities:

- **Continually reinforce the value you place on collaboration.** The overall environment of your center and the climate of individual classes should reflect a spirit of collaboration and sharing, with kids' work hanging on the walls and with chairs placed around tables to encourage interaction.

- **Encourage kids to communicate with one another while they are working on projects.** Model a spirit of inquiry that encourages asking questions.

- **Watch your kids to see who's shy about sharing and encourage those children.** Use the shy kids as assistants and ask them questions.

- **Structure all projects so that kids work in pairs or small teams.** Except for more complicated projects, like creating a video, four people is usually the largest effective size for a team. With larger teams, at least one kid will tend to tune out.

- **Follow a pair-share model whenever you can.** This is especially important when you are introducing new projects and activities and when you reach significant new stages of ongoing projects.

- **Hold brainstorming sessions in the early stages of a new project.** Be sure to encourage all kids to participate, not just those who are most verbal. (See the section on brainstorming techniques, page 94.)

- **Use journals at every opportunity.** If you value journals, kids will become proud of them and look forward to showing off their work. Group journals get kids used to working collectively on even small projects as a matter of habit. Use the pair-share technique with every journal activity. (See the section on using journals, page 99.)

- **Encourage peer experts in your classes.** For example, if kids are working on a photo-editing project and you see that one or two have discovered Photoshop filters or some other advanced feature, ask the class, "How many of you have tried the filters?" When those kids raise their hands, point them out to the group by saying something like "These are your filter experts."

- **Always have children share completed projects with the entire group.** For inquiry-based projects, for example, be sure to include a reporting phase in which kids describe what they've learned.

Techniques for Keeping Children Focused

Children's minds are bound to wander, and all children sometimes act up; these are facts of classroom life. Classroom management is a complex subject and cannot be thoroughly covered in this guide. However, a few pointers for keeping your classes on track are offered below. If you make all of these techniques a steady and consistent part of your classroom behavior, you'll actually train kids over time to be more attentive.

- **Be energetic!** Nothing captures a child's attention like movement and sound, so incorporate both and keep your activity level high.

- **Interact; don't lecture.** Constantly ask questions, even rhetorical and leading ones. In your demonstrations, start questions and let the kids finish them. Show them something, then ask, "Why did I do that?" rather than simply explaining.

- **Bring the kids up close to you.** When you're modeling or demonstrating a concept, they'll feel more involved if they're right near you.

- **Get kids in the habit of learning from each other.** When a child has a question, try passing it off to another student first if you think that student knows the answer.

- **Point instead of calling out names.** If all you do is call out a name when you want a response from a student, the kids don't have to look at you until they hear their name called. If you point and are constantly moving around the room, however, they will have to stay focused on you. It won't seem rude once they get used to it, especially if you sometimes use their names.

- **Don't settle into an obvious pattern.** In an activity in which you're calling on (or pointing to) many people for ideas or suggestions in turn, go back to someone you already called on. Once everyone knows you'll do that, the kids can't tune out after you've called on them once. When you're modeling or demonstrating something with one group, ask a question or otherwise involve someone from another group all of a sudden to keep everyone attentive.

- **If someone's starting to act up, put her in charge of an activity.** She'll have to become the responsible one.

- **Be creative when trying to get the group's attention.** Is it time to call the kids back together after they've been working for a while in teams? Try this: Instead of saying anything, just start snapping your fingers in rhythm. Soon someone will notice and start doing it too, and then the whole class will pick it up. But don't clap; that's too loud and dramatic.

- **Share responsibility with the kids.** Don't always be the one to pick the next person in a series activity; let the kids choose sometimes.

- **Exploit the kids' prior knowledge.** One of the reasons kids sometimes tune out is that their age or the community they come from has given them knowledge and experience different from the teacher's. If you use examples from their world, they'll stay engaged.

- **Use creative ways to manage transitions.** When you're walking with kids from one place to another, have them line up in a different way each time, such as by height, birthday, or alternating boys and girls. Along the way to where you're going, have them count things, like cars or doors.

- **Ignore some kids at times if you need to.** Sometimes it's best not to pay attention to kids who are zoned out or who are acting "too cool for school," especially if dealing with them would disrupt the class even more or when working with groups of 20 or larger. Get these kids back and engaged as soon as possible, however. One good way is to call everyone up for a demo. Even if they hang back for a while, it won't be for long if they see that the rest of the group is having fun without them.

Techniques for Building Reading Skills

An out-of-school program is not usually the primary place where children learn to read, but it can be the place where kids develop a life-long love of reading. It's also where instructors can provide the invaluable service of giving kids extra time and attention in an environment in which there is the flexibility to address kids' individual learning styles. Here are some tips that should help with kids of every age:

- **Read aloud at least once every session.** Use a book, poem, or article that fits thematically with your work for the day. For example, your current project might be building a web page about animals in your neighborhood, using photos you plan to take that day. With a group of early learners, you could read from a picture book about animals; for older kids, you could read from and show the sophisticated photos from a book depicting a famous artist's work. Or you could read part of *Charlotte's Web* or *Frog and Toad*. You don't have to be literal, but make sure there's some connection to the day's work. Don't spend too much time on this activity; it's supposed to energize the kids and prepare them for the day.

- **Mix reading activities into your day**. Make these activities seem regular and natural, not forced or mechanical. You don't simply want kids to read—you want to inspire a love of reading. In the middle of talking about a community project, you might read a short poem that reflects the value of community.

- **Be sure to select age-appropriate materials.** Check the publisher's recommendations, but don't be locked in by them. You know your kids and what's right for their levels of achievement.

- **Take into account the children's prior knowledge.** Choose materials with themes and language that fit with their cultural, educational, and cognitive experiences.

- **Speak clearly when you read aloud.** Don't artificially drag out the words, but do speak slightly more slowly than you would in normal conversation. The kids are trying to follow the story, a process that involves a little more conscious effort than simply holding a conversation.

- **Keep in mind that excitement is more important than theatrics.** Your goal is to have the kids feel and respond to your emotional involvement and the joy you experience in reading, not to do a one-person play. Sure, you should try to reflect the tone or voice of speakers in books, but don't ham it up to the point that your acting overshadows the book.

- **Actively engage the kids in the material.** Try pausing now and then to ask a question about what has happened in the story, or have the kids repeat the refrains from poems.

- **Set aside some silent reading time.** Schedule time for kids to read their own books. During that time, you have to read too. Kids aren't dumb, and they'll know when you're not really into it—so don't try to fool them. Make sure you have something on hand that you really want to read.

Techniques for Brainstorming

One of the best ways to inspire creativity in a group setting is through the use of graphic organizers. Graphic organizers are visual instruments that use simple, repetitive structures to show the relationships between words or ideas. Graphic organizers are simple enough to guide and record the creative process without interfering with the flow of ideas. The aim of using graphic organizers is to break the thinking process down into logical steps so that the group can focus on one issue at a time. They are usually nonlinear, so they can show multiple relationships between words or concepts, and they are well suited to group activities because their simplicity encourages participation.

Graphic organizers are most valuable when you already have a general concept in mind and are ready to search for specific ideas for implementing it. For example, if you know that you want to focus on basic math skills for a particular project, you might use a graphic organizer to lead a discussion with your colleagues, students, or both to come up with project ideas.

Many variations on graphic organizers are available. The following sections discuss two of the most useful ones: mapping and webbing.

Mapping

Mapping is a simple and wonderfully versatile technique that you can use for brainstorming, organizing thoughts, and generating ideas. Mapping can help you select a theme for a project, develop a simple story, or add energy and enthusiasm to any number of group exercises.

Whether you're doing a project with the whole class, breaking up into teams, or working on individual projects, mapping should be a part of almost every group activity. The reason is practical as well as philosophical: If you allow group members to suggest their own ideas and make their own decisions (within the parameters of your educational goals, of course), they will be much more engaged, positive, and enthusiastic than if you make all the decisions yourself and simply distribute assignments. Kids will come up with great ideas that would never occur to you, so use that fact in every way that you can.

The mapping technique described in this section is based on asking a series of questions that elicit thoughts from the group. Figure 7.1 shows a sample template.

The process creates a target-style map in stages. In the center, you start by identifying your project; then, in each larger circle, you record the group's answers to key questions. By the time you finish with the final area, you have a map that tells a number of stories and will help you make a decision about your ultimate goal.

A Mapping Example

Suppose that you've already decided to take two field trips during the upcoming term, but you haven't decided where to go. You gather your group together and use a mapping project to

make the decision. Before you begin, you need to know the following:

- The topic to be mapped (in this case, field trips)
- The goal of the process (in this case, "to decide where to go on a field trip")
- The questions you intend to ask
- Any essential background that will affect the questions you ask or the guidance you give before starting the process (e.g., maybe the semester is dedicated entirely to learning about music).

You also will need the following materials:

- A large pad of white paper at least 2' x 3' (preferable) or an erasable whiteboard or blackboard
- At least four different colors of markers or chalk, suitable for your display board.

Some large projects require additional sheets of paper, preferably large ones like the pad, so that group members or teams can work on individual maps.

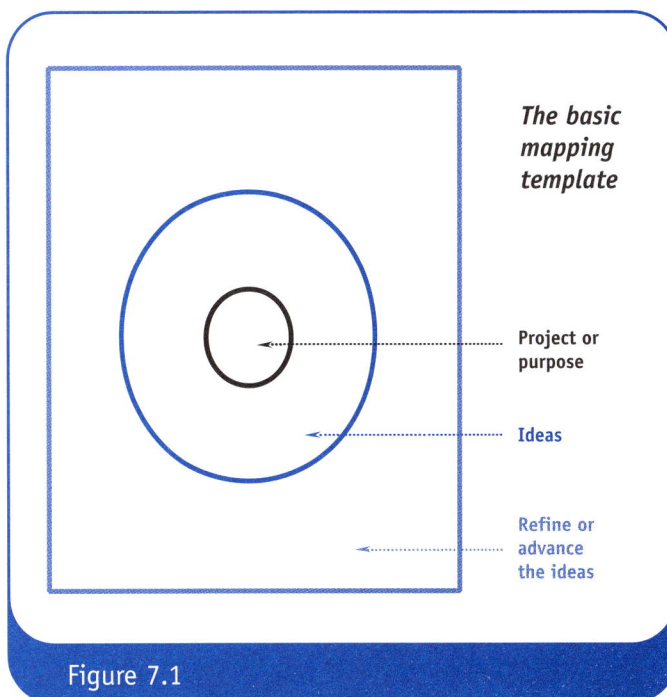

The basic mapping template

Project or purpose

Ideas

Refine or advance the ideas

Figure 7.1

Step 1: Identify the Project or Purpose

As group leader, you'll facilitate the process at the head of the room. Take a marker and write "field trips" in the center of the paper. Now circle it, as shown in Figure 7.2. The center of the map always contains the name of the project or purpose to help keep the group focused.

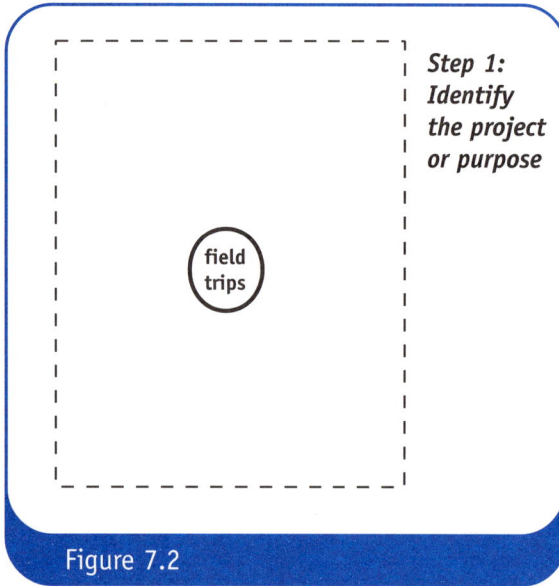

Step 1: Identify the project or purpose

field trips

Figure 7.2

Step 2: Gather Initial Ideas

Take a different color pen and ask the group your first question. Provide any necessary background, but keep the question very simple. If you have a lot of leeway with your curriculum, you might ask, "Where do we want to go?" as shown in Figure 7.3. If you've already decided that the term will focus on music, you might ask, "Where can we go to learn about music?" or "Where can we go to hear music?" Remember, you have to spend time before the session considering your goals and questions because you are looking for focused creativity and input.

You might have to get the ball rolling by making a few suggestions of your own or by picking out people and asking them the question directly. Once things get moving, if someone is consistently quiet, call on him or her specifically. Go ahead and make compliments or jokes as suggestions are made—the exercise should be fun! Just be sure that you're positive and energetic and never judgmental. Even if someone's idea is off base, that's not important now.

You're just brainstorming, and you'll make your decisions later. If you create the right environment, people will be excited because you are asking for their ideas and input.

As kids or group members shout out their ideas, write them on the board around the outside of the center circle. Try not to range too far across the page, because you have at least one more circle to add. (That's another reason to ask focused questions—you want to write just a word or two for each answer, not long phrases.) Move fast and keep talking to maintain everyone's energy level and enthusiasm.

Use your instincts to determine how long to let this round continue. If suggestions are starting to slow or the page is filling up, move on to the next step. Just be careful not to dampen the enthusiasm if things are really clicking. Remember that one goal of this kind of activity is to inspire collaboration, teamwork, and camaraderie. If that's happening, keep writing: It's worth it. Once you decide to end this round, let the group know by saying, "We'll take just a few more ideas."

Graphic organizers work with any age group or skill level. If kids are too young to read, draw a simple picture instead of writing the word. Better yet, do both, especially if part of your goal is to work on elementary language skills.

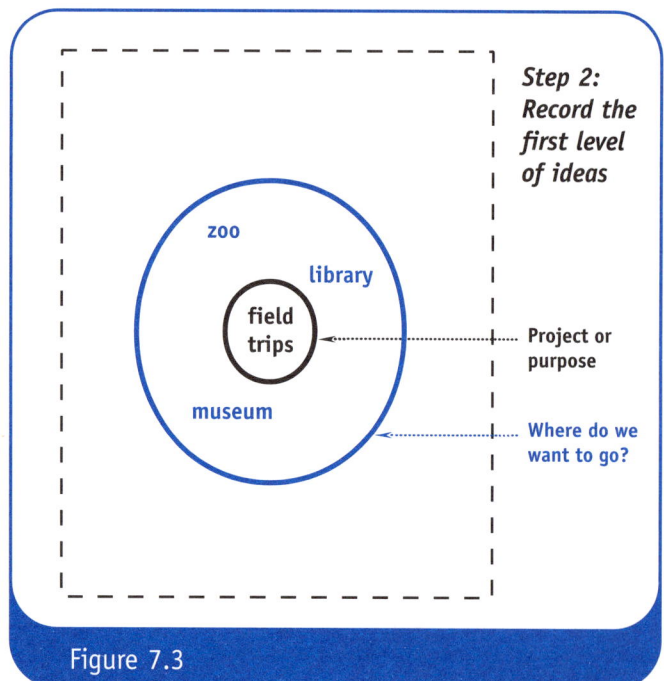

Step 2: Record the first level of ideas

zoo
library
field trips
museum

Project or purpose

Where do we want to go?

Figure 7.3

Step 3: Refine or Advance the Question

Draw a circle around what you've written, and ask your second question. This question can go in many different directions, depending on your goals. As shown in Figure 7.4, you might ask, "What can we do there?"

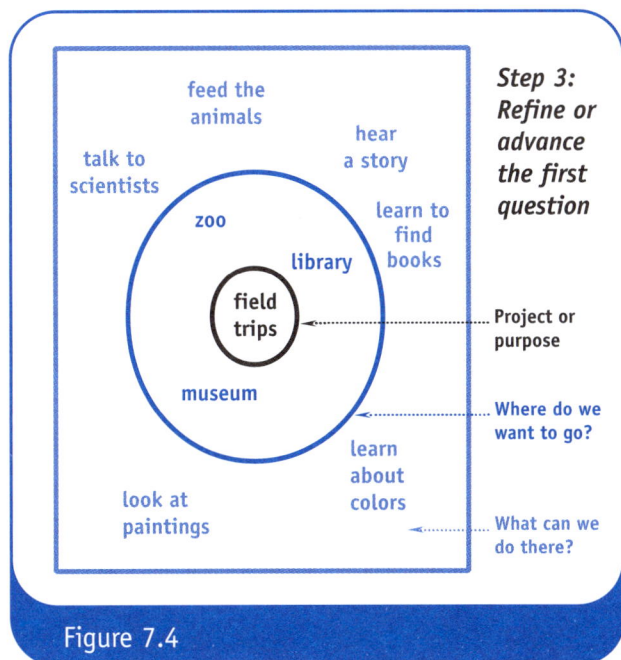

Figure 7.4

Direct the question first to one of the inner ideas by asking, for instance, "What would we do at the museum?" With a different color pen (different colors help distinguish between the levels), write the suggestions outside the new, larger circle. Try to write the answers in the general vicinity of the ideas they relate to. If you're working on a well-focused concept, however, you'll find that many of the suggestions in this circle apply to more than one idea in the middle circle. If so, that's great, and it shows one of the advantages of nonlinear graphic organizers over traditional outlining. Maintain an encouraging tone, and let the group range freely from one of the inner-circle ideas to another. It helps if you can be somewhat orderly by dealing with each inner-circle idea one at a time. But ideas will come when they come, so be flexible. Again, use your instincts to determine when to end this round.

Mapping Results

If you follow the different paths, connecting the ideas from the center circle to the outer circle, you can create logical sentences that tell a story:

On our *field trip,* we could go to the **museum** to *learn about paintings.*

On our *field trip,* we could go to the **museum** to *learn about colors.*

On our *field trip,* we could go to the **library** to *learn about colors.*

On our *field trip,* we could go to the **library** to *learn about books.*

On our *field trip,* we could go to the **library** to *hear a story.*

On our *field trip,* we could go to the **zoo** to *hear a story.*

On our *field trip,* we could go to the **zoo** to *feed the animals.*

Returning to the goal of the exercise (Which field trip do we want to take and why?), you could almost decide at this point, or the group might revisit each of the ideas in more depth. For example, you could create follow-up maps dedicated to each place and ask additional questions to continue to refine the topic. If you were trying to plan an entire project for which the field trip was just one activity, you might go on to a new map to answer, "How can we tell people what we saw at the zoo?"

Webbing

Graphic-organizing techniques that build connections between similar words or ideas are referred to by a variety of names. We will use the terms "webbing" and "web maps." Webbing works best when you want to show a lot of words or ideas and keep them loosely connected.

The Structure of Webbing

Start by thinking of the kind of map described in the previous section. Suppose, for example, that you're working on an activity about computers. You might come up with a map that looks like the one in Figure 7.5.

In this map, the core topic is "computer," the middle circle is "parts of the computer," and the outer circle is "what the parts do." It could form the basis of a sentence pattern, such as "I have a computer with a keyboard that lets me type." With this pattern, however, you might be more interested in coming up with lots of different parts, and you also might be interested in the direct relationship between the computer and its parts.

A traditional outline might represent those connections this way:

I. Computer

    A. Keyboard

       1. Type with it

       2. Special keys

       3. On/off button

    B. Monitor

       1. Shows pictures

       2. Shows icons on toolbar

       3. Shows lots of colors

Although outlining is good for composing reports or long text documents, it is so highly structured that it doesn't work well for coming up with ideas quickly with young kids.

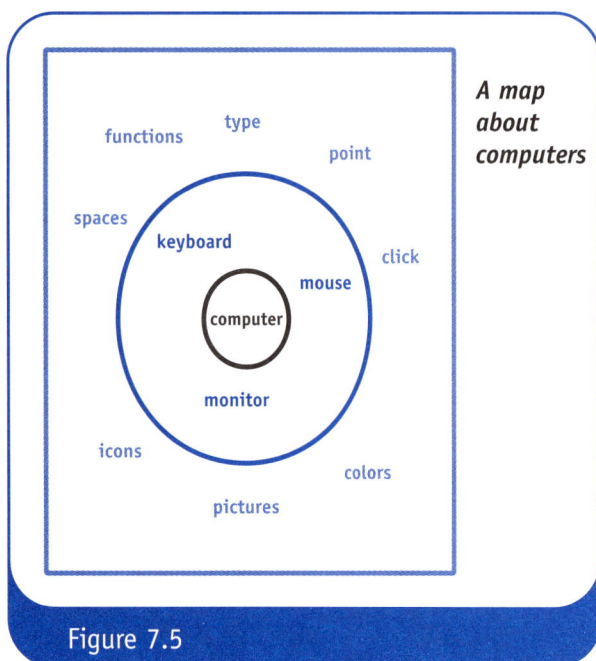

A map about computers

Figure 7.5

Figure 7.6 shows a simple web map. To create it, you write the topic in the center and circle it, just as with the first map. You then write another word, circle that one, and draw a line between them.

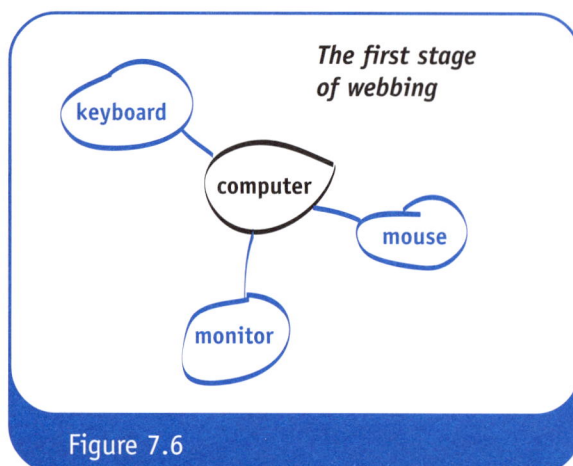

The first stage of webbing

Figure 7.6

As you can see, webbing is easy with only two levels but gets a little trickier with three (see Figure 7.7). The method still works, but you have to leave lots of room between your first-level words. If you have more than three levels, webbing usually won't work because of space problems; you're better off creating a new map for each third-level word that you want to expand on. You can see that webbing has the advantage of showing the direct relationship between words, making it an excellent tool for pattern-writing exercises. (Pattern-writing exercises are activities based on identifying and repeating sentence patterns. They are particularly useful for teaching writing, sentence structure, and parts of speech.)

Web maps can be extremely useful for other projects as well. Try webbing for brainstorming activities in which you want to collect many ideas quickly and maintain a way of showing relationships. For example, suppose that you want to do an interviewing project with the class. Before having the students make up a list of questions for the interview, get their big-picture thinking organized by using a web map to figure out the topics for questions first. Then the class can come up with specific questions around each topic.

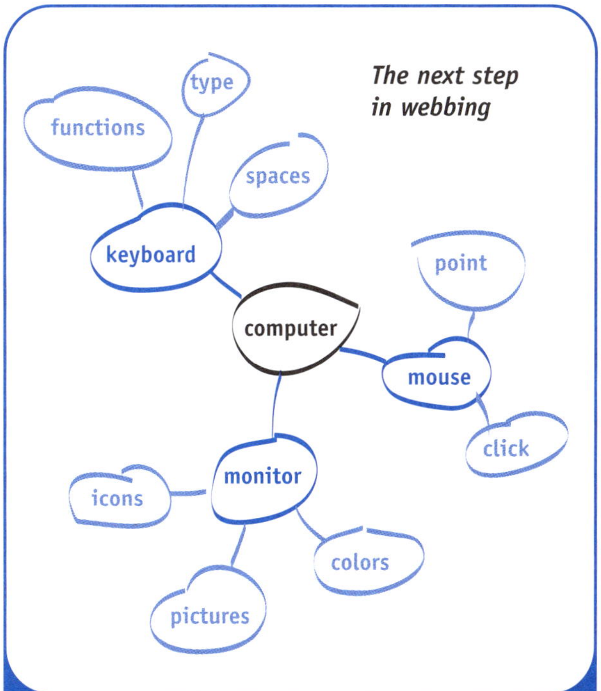

Figure 7.7

The next step in webbing

Webbing on the Computer

A number of computer software programs are available to help kids organize their thoughts. For example, Inspiration offers a variety of tools and formats, including webbing options; a version called Kidspiration is specifically designed for young children.

Techniques for Using Journals

On the first day of the term or program session, give each person a blank book to keep as a journal. Have the kids make the books their own by writing "My Journal" on the cover, or something similar, along with their name, the name of your center, and anything else you (or they) would like. A journal is much more than a diary. In fact, it's important that kids not think of it as a diary, because although a journal is personal, it is not private. One of the important uses of the journal will be to encourage sharing and communication by having kids show their work to a partner and to the group.

But if it's not a diary, what is a journal? It's more like an "everything" book. It's a place where kids can write their thoughts and ideas, but it's also a multimedia project where they can draw and paste things they find or make, whether it's a photo, a leaf, or a souvenir from a field trip. Think of it as the "default" medium for your activities so that the kids become so attached to their journals that they carry and use them wherever they go. Whenever they have an idea they want to follow up on or a thought they want to remember, whenever they are feeling creative and want to doodle or write a poem, you want them to turn to their journal as a trusted friend and a safe environment. In order to achieve that, you'll have to do two things:

1. Integrate journals regularly into any activity where it feels appropriate. You might want to begin each session with a short journal activity that anticipates your theme or a lesson you'll be working on. Such activities should be energizers to start creativity flowing. What's more, if every day you have the assignment written where everyone can see it, the kids will learn to start working on it as soon as they come into the room. It's a great way to keep them productively occupied while everyone filters in.

2. Keep a journal of your own that you frequently share with the children in your program. In the spirit of good modeling, it has to be a real journal, one that you truly care about and maintain outside the program. If the children feel your pride and sense of enjoyment in your journal, they'll emulate your behavior.

What kind of book should you use? The only real requirement is for it to be of a standard notebook size that gives plenty of room to work on each page. The traditional school composition books (with the black-and-white covers and tape bindings) serve the purpose well for two reasons. First, they are inexpensive. Second, they feel like a real book, not just a source of paper. The kids won't be tempted to tear out a page every time they feel they've made some small mistake or if they need a scrap of paper, the way they do with spiral notebooks.

It is also appropriate to have the kids make their own books with blank paper bound simply between cardboard covers. Use lined paper that provides a loose yet nonintrusive structure for certain kinds of writing and drawing activities. In addition, remember to keep a supply of creative materials readily available so kids can

7

Develop Effective Teaching Techniques

©2001 Education Development Center, Inc. All rights reserved.

99

have fun with their journal activities, including pens and markers of various colors, crayons, paste, glitter, and sequins.

Here are some other ideas for using journals in your program:

- **Use journals for graphic organizers.** Whenever a project or activity has a graphic-organizing component to generate or organize ideas, have the kids do their personal maps in their journals. Also, encourage the use of journals for brainstorming activities.

- **Use journals for outlining and making notes.** If you're doing a project in which kids will have to come up with many ideas—an interviewing project, for example, in which you'll want them to come up with lots of questions—have them write lists of ideas and questions in their journals.

- **Have kids record new knowledge.** Provide kids with a selection of websites on a particular topic. Ask them to visit one or more of the sites and write in their journals three things they learned. Then have the group members share and discuss what they wrote.

- **Use journals for nonwriting activities.** When teaching drawing or doing drawing practice, have the kids do their work in their journals rather than on loose sheets of paper. Any time you create other projects that will fit into a journal, such as drawings and printouts of computer work, have the kids tape them in. Try using reusable tape so that they can be removed and replaced easily.

- **Do pattern-writing activities.** As you read sentences aloud, have the kids write them down in their journals, or provide sentence templates for which they can write their own variations in their journals. If you're doing a group activity, such as creating poems based on a pattern, have the kids record the finished versions in their journals when the group work is done.

- **Schedule writing time, especially with older children.** Have them spend 5 to 10 minutes or more doing "free writing" in their journals on topics you select or that

they pick themselves. At the end of the day, have kids spend a few minutes doing evaluative projects, such as recording in their journal "What we did today" or "Three things I learned today."

- **Have kids create a list of technical tips.** When kids learn a new technique at the computer, especially ones that are a little complicated or have several steps, have them record it for future reference (e.g., "Here's how I check my email" or "How to make a web page").

How to Use Journals Outside Your Program

One of the many reasons for relying on journals in class is to encourage kids to make their journals an important part of their lives outside your program as well. Here are some ideas:

- **Give kids simple journal assignments**. For example, have them describe people they meet or things they'd like to learn. Be sure to make the assignments fun and not at all like homework. Kids should want to use their journals, not see them as punishment or extra work. Be careful never to confuse the journal with a workbook or only a place to take class notes. Journals should feel personal and special.

- **Encourage kids to carry their journals around with them.** They can record their thoughts, ideas, or feelings when riding the bus, sitting in the library, or anywhere else.

- **Make journals an important part of field trips.** Use them for everything from pasting in a map before they leave, to recording thoughts and observations while they're on-site, to writing summaries of what they learned.

- **Meet with children individually to review and discuss their journals.** Try to do this once a week, if possible.

Other Kinds of Journals

You may want to integrate some other kinds of journals into your program.

- **Group journals** add a collaborative as well as a personal component to activities. A group journal is just like an individual journal except that several people work in them simultaneously. For example, you might have your room set up so that you have a group of four kids around each table. You could give a journal to each group and have the kids work on a single mapping exercise together. You could also give a group journal assignment to a team of kids working on a longer inquiry-based learning project.

- **Dialogue journals** offer an opportunity for you to provide more private and personal guidance to students, especially when teaching language and literacy. A dialogue journal is a written conversation in which a student and teacher communicate regularly, responding to questions and comments, introducing new topics, and asking questions.

- **Online journals** bring computers and new media into the world of journals. The Diary Project (www.diaryproject.com), for example, encourages teens to write about their day-to-day experiences growing up and share their writing with others. If you're feeling ambitious, you can try doing something similar for your program or center, or simply have kids transfer some of the entries from their journals to web pages that others can view online.

Techniques for Teaching About Technology

Working with personal computers (PCs) and network technologies may be new to you or others at your center. Even if you've used a PC, word processor, or the Internet, you may not feel that you know enough to teach kids how to use them (or to use them as teaching tools).

Don't worry. No one expects you to become a technologist, nor should you expect to have all the answers. It is the instructors' teaching skills that are most important; the specific technology knowledge can be developed along the way.

Instructors and aides will need some basic introduction, of course, to any hardware devices or software applications they intend to use in class. Their knowledge doesn't have to be extensive, however. It's okay to learn along with the class. If the instructors are unfamiliar with a piece of software they want to use in a project, make sure that they get an introduction to its basic concepts, whether from some of the resources listed throughout this manual, from a colleague, or from a trainer brought in by your center. Below are some general tips to help you use technology effectively in the classroom.

- **Integrate technology into your larger goals.** Learning how to design web pages or use drawing programs should be part of a project on building language or math skills, not an end in itself. Kids need to understand that technology is just one of many tools for learning and communication, like a book or a pencil. In fact, one of the most important things you can teach kids is how to decide when one tool or medium is better than another. For example, a handwritten letter is sometimes much better than email.

- **Work up to more advanced technologies from simpler ones.** For example, suppose you are doing a project in which you want children to draw a picture on the computer and ultimately add it to a web page. Long before you introduce a drawing application like KidPix, you should teach the fundamentals of drawing with crayons and paper.

- **Assess how much the kids already know.** Especially when it comes to the Internet, it's hard to guess what they know. You'll probably find big differences among the children, just as with writing and reading skills. One child's parents may have an Internet connection at home; another child may go to a school with a computer lab more advanced than your program's.

- **Use peer teaching between kids.** When you do find kids who know a little more, are a bit more adventurous, or are more engaged by a technology, use them. After all, it is guaranteed that you are not going to know everything about every piece of software or hardware. The way kids relate to other kids is different from the way they relate to you, and in some cases they will learn more effectively from one another.

- **Keep it simple when choosing software.** Remember, having a large number of features does not always make a software application a better tool, and most people use only a small percentage of the features in a software package. (See page 112 for more information about choosing software.)

- **Don't sweat the small things.** Things will go wrong. One day you won't be able to remember exactly how to find an option, or the network connections will have problems, or something won't work in your file even though it did yesterday—it's just part of the world of technology. Laugh about it, work around it, and keep moving. It's going to happen to the kids, too, so it's important that you model the proper way to deal with challenges. Make mistakes on purpose in your demonstrations occasionally, just so you can show the students that it's not a big deal.

- **Don't explain things too deeply.** The less you say, the more your students will explore for themselves. Remember, your goal is not to create software experts, it's to teach core curriculum skills and help kids understand the many communication, research, and creativity tools they have at their disposal. Good critical thinking is more important than technological proficiency. When a child asks, "How did you do that?" see whether any of the others can answer before you try to explain it.

- **Take your time.** When introducing new technologies, do it in the context of larger projects that extend over many sessions. In this way, you can first introduce the conceptual skills (e.g., drawing on paper), then slowly introduce each piece of equipment or software. Students will benefit from a progressive introduction to how to use the tools, and repetition builds support for their understanding.

Email and Online Communities

The Internet isn't really about computers; it's about people communicating and collaborating. Even though the web seems to get all the media attention, email and other messaging applications may be the most powerful Internet tools of all. They connect people directly, just like the telephone, with advantages the telephone lacks, such as the ability to communicate with many people at once at their convenience.

If you really want to help kids become successful in the digital age, focus less on the tools and more on the human communication skills they'll need to use those tools effectively. After all, email is about more than sending a message: It's about the message being sent, who's receiving it, what the message can do, and what response is desired.

"Messaging" refers to the use of any software program that lets you send a message directly to one person or to many people at once. Messaging differs from web use in that it involves your taking the initiative to communicate with a specific person or to react to a message sent to you. Websites, in contrast, passively wait for someone else to take the initiative by visiting them. Moreover, website visitors are anonymous, for the most part, unless they choose to identify themselves by sending a message to the site owner.

Email

Email is by far the most important and popular messaging application. The number of email messages sent each day far surpasses the number of letters sent through the post office. The convenience and versatility of email have made it an essential tool for doing business today in every field.

All email programs are simple to use, and it won't take long to present them to your kids. Follow the guidelines for teaching about technology described in the previous section when introducing the software. It is important that you train kids to think about communication, which means that you need to do two things:

1. Use email frequently to reinforce basic communication and language skills, such as reading and writing. Try activities in which you send email messages to each other, to kids in other classes, and to other people in the center or community. Try finding email "pen pals" in other parts of the country or the world. In addition, you should familiarize the children with some conventions of communicating through email, called netiquette.

2. More important, train kids how to communicate with others thoughtfully. Email is a great tool for doing this because it lets people respond at their convenience. Teach kids to contact experts and sources (politely) through email for their research projects and to build communication features into their web pages. Even just putting their email addresses on printed documents helps them internalize this habit.

Don't forget to teach kids about safety on the Internet. The Internet is just a reflection of the larger world, and some predators look for kids through email and other messaging applications. (See the section on online safety in Chapter 8, page 119.)

Online Communities

The next step in using email is to introduce communications that involve groups of people. The following tools can be used for "many-to-many" communications:

- Electronic mailing lists allow many people to exchange information and/or have a discussion through email. Electronic mailing lists focus on a topic; the topic can be as big as "technology and society" or as specific as "lesson plans using KidPix for young children."

- News lists, such as electronic newsletters, generally are one-way broadcasts to people who are interested in a particular subject. Usually no communication occurs among the subscribers, although people may sometimes communicate back to the editor, much as they would write a letter to the editor of a magazine.

- Bulletin boards, forums, and threaded discussions are web-based facilities that allow you to post a message for people to read at their convenience. For example, you could create a bulletin board with other youth centers or nonprofit groups in your area to post messages about sources for supplies that you can share or exchange. The main difference between an electronic mailing list and a bulletin board is that people receive list postings right in their email inboxes. Bulletin board formats require users to "go" to them—to get on the web, type in an address, and read each message.

Web Pages

If you look at a web page in its basic form, what do you see? Text and pictures. When you work with web pages, don't worry about all the bells, whistles, and fancy technology; instead, focus on the basics of good communication: text and pictures.

Web pages are easy to build, but make sure that children are ready conceptually before you start teaching them how to do it.

Children should first be introduced to the web itself, of course, and email. You should spend time introducing topics like digital photography, image editing, file formats, and multimedia concepts before creating a web page. That way, students will learn good visual and written communication skills as well as technology.

Build a website for your center and include areas for pages from different classes and programs as well as from the individual students. Create a way for students to show off their work online. Even if you can't make the site available over the Internet, you can still put web pages on your lab computers; people can see them when they visit your center.

What You'll Need to Build a Web Page

Keep in mind that web publishing is really a multimedia exercise. Web publishing software is simply used to enter text and assemble objects on a page. If you have already introduced the kids to word processing, drawing, image editing, or multimedia software, they'll already know much of what they need.

You have many choices in web publishing software. We recommend Netscape's Composer because it is easy to use and has the advantage of being free and available on most community computers. As a result, kids can take their files home or continue their work at a library or school. Another good web publishing product is Macromedia's Dreamweaver.

If you plan to connect your pages to the Internet, you'll also need to work out the details with your network administrator or other technical staff member, although you may not have to concern your kids with this information.

Web pages are actually files written in a programming language called HTML (or one of its variants). A web publishing application lets you lay out a page visually so that you don't have to learn programming. You simply type in text and arrange your links, pictures, and other objects on the page; the program automatically converts everything to HTML. It's easy because the program does all the technical work. For older kids, especially if part of your goal is to help them gain career skills, you may want to bring in local professionals from Internet design firms or technology companies who can offer more detailed information about HTML and professional web development.

A Structure for Teaching About Web Pages

Because you can take so many routes to web page creation, it is important to remember the basics:

- **Make web work part of a larger project.** A web page can be the culmination of almost any kind of project, from a long-term inquiry-based project to a field trip.

- **Have kids do sketches in their journals first.** This will help them think through what they want to do before they start building an actual web page. Another reason for this step is to show them that new technologies augment, rather than eliminate, traditional methods.

- **Be sure to talk about the features that make a good website.** Talk especially about the importance of including interactivity and communication to web pages, such as including an email address so that people viewing the students' web pages can respond and give feedback. Kids may not think about these elements at first, but talking about the subject will help them internalize the habits of online communication and will positively reinforce their work. Just remember that adding the email address is not enough by itself; you'll need to let people know that the page is available, even if it's just parents or kids in other classes at your center.

> @ *For more information, examples, and resources, see the extensive YouthLearn website at* www.youthlearn.org

WORK WITH TECHNOLOGY:
chapter 8 **THE NUTS AND BOLTS**

8

Overview

As you have seen, this manual focuses on learning first, not technology, in the belief that learning is the primary objective and that technology is a useful tool to enhance that learning. But now you're ready to focus on getting the technology components in place. This chapter will help you consider carefully the mechanical aspects of developing and running your technology learning center. For those of you who already have a basic understanding of the workings of technology, some sections of this chapter will be "old hat"; feel free to refer only to those areas in which you want or need some additional guidance.

The Physical Space

When you create a learning environment for kids that incorporates technology, a number of important issues must be considered. A good design helps kids make the most of the computer equipment, allows instructors to manage the class effectively, and provides a setting that is welcoming, safe, and fun.

The design of your center will depend on the number of people you intend to serve, the kinds of activities you plan to undertake, and the age range of participants. The following sections offer guidelines and questions to ask yourself along the way.

The Room

A well-lit room with free wall space and easy access to a nearby secure closet is the best environment for a technology learning center. Ideally, the space should be large enough to accommodate not only the technology-related furnishings but also worktables, shelving, and space for activities that don't involve technology, such as arts and crafts, reading, and group discussions.

The Layout

Arranging the machines around the perimeter or in inward-facing islands allows instructors to move easily around the room and offer assistance. Layouts that foster collaboration and accommodate offline activities are particularly important if you are trying to incorporate inquiry-based learning.

Collaborative learning is very effective in a youth technology center. Plan to place two students at each computer occasionally, if not frequently. Space the computers far enough apart to make this possible.

Contents

- Advice on setting up your center's physical space and selecting hardware and software
- Tips for enhancing youth safety online
- Advice on technical support providers

Audience

- Program directors
- Organizational leaders
- Educational administrators

Remember to include additional space for peripheral equipment, such as scanners, cameras, audio or video devices, or large monitors. A lockable cabinet to store portable equipment, such as digital cameras, will reduce the chance of theft.

Space for activities that do not use the computer (e.g., reading aloud and drawing) ideally should be part of the computer lab space. An integrated space makes it possible to conduct different kinds of activities without having to move the class physically into another room. If that is not feasible, create an activity space as close as possible to the computer lab.

The Furniture

Many ergonomic computer products have been designed specifically for children. Seek advice from a reputable dealer before making a purchase.

Children with physical disabilities may require furniture or equipment designed specifically for them (e.g., special tables that can easily accommodate wheelchairs, extra-large terminals, Braille keyboards, or special audio devices). Make sure your center and its equipment are accessible to all participants in your program.

Kids of very different ages and sizes often share the same center, making it difficult to design an ergonomically correct space in which users are seated at the proper height and distance from keyboards, mouses, and monitors. Chairs with adjustable height solve only part of the problem, because seven-year-olds will have to dangle their feet uncomfortably if chairs are raised to be closer to a too-high table. Ideally, the chair, the height and angle of the monitor, and the work surface should all be easily adjustable. If that is not possible, provide some workspaces with chairs and tables that are meant for smaller children.

Allow at least 18 inches between keyboards to give students room for books, paper, and other supplies.

Maintenance

Avoid direct, unfiltered sunlight on equipment. If possible, try to get a northern light exposure so that direct sunlight is minimized, or install blinds that can be partially closed during the day.

Air conditioning is essential to reduce the heat generated by computers. A computer lab can generate so much heat that the room may become unhealthy for people and damage the machines. Be sure there is sufficient air conditioning and ventilation to operate the computers during warm weather.

Careful and regular vacuuming of the carpeting is absolutely necessary for proper maintenance of computer hardware. In addition, at least twice a year the machines should be cleaned and vacuumed, with covers carefully removed and then replaced. Some technicians use compressed air (found in computer-supply or electronics outlets) to clean the interiors of equipment.

A moistened rag can be used to clean the outside of the computer case. Keyboards need special attention because they are notorious dust collectors. They should be wiped down regularly, and great care should be taken not to introduce moisture into the keyboard mechanisms.

Never use the computers during storms or if a storm is kicking up. In the event of a major power surge you may ruin the computers beyond repair. And never let kids (or adults, for that matter) bring drinks or food into the computer lab.

Safety Guidelines

Children, especially those with mobility impairments or disabilities, should have safe passage to and from the computer lab. Note the following safety guidelines:

- Secure conduits and wire bundles so that people don't trip. Raised floors and drop ceilings, which allow for direct cable connections to work areas, are preferable to cable covers, conduits, and tracks that cross open floor spaces.

- If a cable must be run across an open floor area to reach a centrally located island or station, make sure the cable covers can't be dislodged, kicked, or moved, even by a relatively purposeful kick.

- Electrical codes should be carefully adhered to in order to ensure safe operation and years of reliable use.

- Uninterruptible power supply units, if used on a per-workstation basis, should be hidden if at all possible. Many of these systems contain corrosive acids and other toxic chemicals.

- Children should be informed of the electrical hazards around them and instructed not to touch the power supply under any circumstances.

The Network Closet

The network closet should be a neat, spare room used primarily to house the server (computer processors that control your network) and not for other purposes. The room should be cool and dry, without nearby plumbing or any chance of leaks, moisture, or excessive heat. The closet should be secure, with a solid door and a reliable lock in place to prevent unauthorized access. Only people who are in a position of trust should have the key or know the combination; at least one such staff member should be physically present at the center on a regular basis in order to provide access to the closet, if needed.

Equipment Inventory and Security

Keep in mind the following tips on equipment inventory and security:

- Inventory equipment when it is purchased or donated, and keep the equipment in a secure area until it is installed. Write down the serial number, make, and brand of each piece of equipment.

- Shipping labels, invoices, packing slips, and other documents should be carefully examined and stored in a safe place, such as a file cabinet. The labels and invoices will be invaluable if you ever need to repair or replace equipment.

- Keep the system documentation, manuals, installation guides, and other technical literature in a centralized location.

- Unobtrusive locks that attach desktop computers to tables will discourage anyone

who might want to remove the equipment without authorization.

- It is a good idea to provide labels or signs to indicate who is responsible for equipment maintenance or problems and what number to call for help. Doing so is particularly important if machines are maintained by outside vendors or volunteers, if personnel frequently change, or if different people share responsibility for a facility.

- Make certain that the facility's property insurance will cover replacement of all the equipment in the learning center.

The Technology Plan

Setting up a technology learning center is a big job involving a wide range of technical issues. Every center is different, requiring different services, equipment, and providers. Before you purchase one piece of software or upgrade a single computer, you need to complete a technology planning process and develop a detailed technology plan.

The technology planning process will help you use staff time efficiently, minimize technology-related crises, and avoid wasting money on equipment that makes your life miserable. A technology plan will help you think through your priorities in order to use technology in a way that directly furthers your mission. It will help you budget for technology purchases. Last but not least, you can use a technology plan as a tool to advocate for technology funding.

The following sections provide a brief outline of the technology planning process. For more detailed information and guidance, we recommend that you visit CompuMentor's TechSoup website (www.techsoup.org).[9]

What Is a Technology Plan?

A technology plan is a working document that describes your current technology and computer status, the upgrades that are needed, the issues that must be addressed, and the budget implications. It is a cross between a funding proposal and a map of your organization's technology. Your technology plan addresses not only equipment—computers, printers, and so on—but also issues such as staff training, staffing and consultants for technical support, your budget, and, most important, the mission of your organization. Your technology plan should initially be developed through an extensive planning process; it should be updated periodically as your programs develop and your technology gets old. Most technology plans include the following sections:

- Your organization's mission

- Your technology goals and vision

- An assessment and documentation of all of your current hardware, software, and information systems

- A plan for staff development and training

- A technology implementation plan with timelines and benchmarks broken down into smaller individual projects

- A budget that includes all hardware, software, system upgrade, and staff costs.

How to Develop a Technology Plan

The first step is to create a technology team— a group of people who will participate in the technology planning process. It is important that this team include more than just the "techies." Your team should include instructors, managers, program developers, and others. It is best to involve from the beginning any "naysayers"—that is, people in the organization who resist spending any money on technology. Once the team is in place, the following steps should lead you to a solid technology plan:

- Analyze your organization in terms of its goals and needs.

- Assess your current technology status, and document existing systems.

- Clarify specific technology goals and strategies. Research the best solutions to your technology issues.

- Establish technology priorities that include your programs, mission, financial status, culture, and current infrastructure.

- Identify training and technical support needs, and create policies and procedures.

- Prepare budgets, funding plans, and grant proposals.

The next sections of this manual are intended to help you think through some of the specific issues you will need to address when writing the technology plan for your center.

Technical Issues

Setting up a technology learning center is a big job involving a wide range of technical issues. Every center will be different, requiring different services, equipment, and providers, so you'll need someone very experienced to help you develop a sound technology plan.

Although it is beyond the scope of this manual to provide all the particulars, below are some general principles to keep in mind.

- Hire a technology consultant to help with your budgeting and big decisions, such as the specifications for your computers and telecommunications requirements. (See the section on choosing a technology consultant, page 118.)

- Budget for a technical support provider. Even if things don't go wrong—but they will—you'll have questions over time that will need answers. Don't skimp here. You'll also need a service contract for your equipment that takes into account your unique needs.

- Designate someone on staff who is knowledgeable about technical issues to serve as your network administrator and technical troubleshooter. The network administrator should be able to manage the system, fix common problems, and help train other staff members in how to use the technology. This can be a dedicated position or one filled by someone who doubles as an instructor.

Make Your Center Internet-Ready

To get on the Internet, you first need computers, of course, so take a look at the section on computer hardware (page 110) to help you work with your advisors. To use the Internet (at least, to use it in classes), you will need four components in addition to a computer:

1. **A way for the computers to connect to the Internet.** An Internet connection requires having either a modem (a device that uses a telephone line to dial into a network) in each computer or a local area network (LAN) that connects all your computers and gives them access to the Internet. Most technology learning centers will end up needing a LAN, unless you have five or fewer computers.

2. **Electric and network wiring.** The physical infrastructure of your center may make many of your technology decisions for you. For example, older buildings may lack enough wall outlets for the computers or the correct telephone lines for certain telecommunications services. Imagine ordering 20 computers, only to find that the building's electrical system won't support them. Your advisors should be able to help you figure out the details. Be sure to include a thorough site review as part of your planning checklist.

3. **A telecommunications service provider.** Your needs for communications services are going to be a lot more complicated than those of a simple residence. You'll need multiple phone lines and, perhaps, integrated data and voice services. Technology is changing rapidly, so consult with your advisors.

4. **An Internet service provider (ISP).** Your ISP may end up being the same company as your telecommunications provider because most providers now offer ISP services as well. An ISP is the company that actually connects your computers to the Internet. Remember that the Internet is really a "network of networks." ISPs are connected directly to the Internet "backbone," and you connect by first connecting to the ISP. Part of choosing an ISP is assessing the services you want. For example, you may simply want the basic ability to surf the web and get email, or you may want additional services, such as hosting your website or video conferencing. Another consideration, one that will affect whether your ISP and telecommunications provider are the same company, is the connection speed you want—faster is better, but it is also more expensive.

The world of Internet connectivity offers an abundance of options.[10] Making good decisions can be hard: It requires a combination of pragmatism and forward thinking. Here are some questions to consider:

- How many users will you be serving per day, per week, and per month?

- What are your priorities? To train people to use traditional software? To produce multimedia projects? To involve kids in project-based learning?

- Will you run a server or network in-house? This is an important consideration if you wish to reduce costs.

- What is your budget?

Many organizations attempt to tackle the last item first, then move on to the other considerations. Truly, budget should be your last consideration. You always want to explore all options before assigning a price tag. Your primary considerations should be who your clients are, how often they will need access, and the internal capacity you have for maintenance (i.e., technical support and so forth).

Email and Email Addresses

You need an email account in order to send and receive email; that's usually a basic service offered by an ISP. Most likely, you will want a separate email account for each staff member. Once you start adding separate accounts for dozens or even hundreds of kids, however, it may be too much to handle. But web-based and free email services are available. (Note that few things are truly free, and most "free" services may append ads to the bottom of your emails; however, they are fairly unobtrusive and are a small price to pay for getting your students connected to the world.)

If you get service through a web-based free service or purchase it through a company, your email addresses will use the company's or service's domain name. For example, everyone who uses America Online has "@aol.com" as the second part of his or her email address. If your center has purchased a domain name for its website, that also may be the domain name for your email addresses.

Keep in mind that a person can have several different email accounts. You might have a personal account with America Online or Hotmail and another at your center. Your kids may have accounts at home, at school, and at your center as well.

Hardware, Accessories, and Peripherals

So many options, combinations, and applications are available in computers today that we can't be very specific in making recommendations. Besides, the speed of change in technology, systems needs, and prices means that the best choice for today may not be best for tomorrow. This section defines some basic terms and lists some general issues to consider in choosing hardware for your center.

Each workstation will need a number of different components:

- **The computer itself.** The main processing chip in the machine, the CPU, is the heart of the computer; the faster the chip, the faster your machines will run. Computer speed is measured in megahertz (MHz).

- **A keyboard and a mouse.** Some systems do not include these essential input devices in their basic price.

- **A monitor.** A wide variety of sizes and resolutions are available. ("Resolution" refers to the detail and quality of the images the monitor can display.)

- **Memory.** The computer uses memory to perform calculations and operations while you are using an application. The more memory you have, the better, especially if you will be using memory-intensive applications like Photoshop or working with video. Additional memory can be added to most computers, but buy as much as you can afford at the outset. As a general rule, getting more memory will often be the most cost-effective way to increase the capacity of your computers. (Check with your technology consultant, but as a general baseline, computers using the Windows 95/98 operating system need at least 64 megabytes (MB) of memory, with 128 MB

recommended. Computers using Windows ME/2000 need at least 128 MB, with 256 MB recommended.)

- **A CD-ROM drive.** Formerly an add-on, CD-ROM drives now come with virtually every computer.

- **Storage devices.** To store your files, you will absolutely want a hard drive of at least 10 gigabytes in every machine. The hard drive is where you will store the applications you use on the computer, most of the files you and your kids are actively working on, and, possibly, archives of old files. Hard drives are the most efficient way to store data and applications so that the computer can access them quickly.

- **A 3.5" floppy drive.** Because floppy disks are the easiest way to transfer small and average-size files other than over a network, you'll want a floppy drive in each machine. If you have floppy disks, you can transfer files from a digital camera and let kids take their files home or to school to continue work. Keep many disks around as part of your store of supplies. Virtually all Windows-based computers come with a floppy drive; new Macintosh computers do not, however, and you must buy them separately.

- **A modem to connect to the Internet** (unless you have created a LAN for your center).

In addition to the components above, you may want to add some of the following accessories to one or more of the computers in your center:

- **A Zip drive.** A floppy disk holds 1.4 megabytes of information, but a Zip drive holds either 100 or 250 megabytes, which is needed to store large graphics or sound files. Zip drives offer a convenient way to store and transfer great masses of data.

- **A DVD drive.** This type of drive is used largely for playing videos and is becoming more and more common on new machines.

- **A CD-RW drive.** To create your own CD-ROMs, you'll need a CD-RW drive. They are not really necessary for the average lab, but if you have the opportunity to acquire them without making other sacrifices, you may want to consider doing so.

Other devices to consider include external speakers, which add quality or volume to audio from the computer; additional storage devices; additional memory; and expansion cards, which can help with specific applications like video.

All computers should be plugged into a reliable power supply because electrical surges, brownouts, and blackouts can damage computers and cause you to lose files. At the very least, a surge suppressor is needed, but an uninterruptible power device is much better because it will continue supplying power to the computers for a brief period after the power goes out, allowing you time to save files and turn off the computers. The warranty on any such device is a good indicator of its quality.

Printers and Output Devices

Your computer lab will probably need at least two printers:

1. A sturdy, fast, high-volume, black-and-white printer you can use for text, drafts, tests, and large print jobs.

2. An inexpensive color inkjet printer. Don't forget to factor in the cost of ink cartridges, which varies depending on the printer you choose.

Consultants can help you figure out the best way to connect the printers. For simplicity's sake, you may want to have them connected to just one main computer, or you may want to have them attached to the LAN (if your center has one) so that all computers can use them. Or, if you want to connect multiple computers to one or more printers without a LAN, you can do so through switch boxes.

Remember that you'll need cables to connect the printers to the computers or LAN. Most printers do not come with cables, and you have to be sure to have the right cables for your printers and their connections to your computers and switch boxes.

The other output device you may want to consider is a projecting device that can be connected to a computer in the lab or to laptops brought in by others. This device will allow you to project whatever is on your monitor onto a screen or wall for demonstrations. Projectors tend to be expensive, but they can be a big help when doing presentations to larger groups; also, because they're portable, you can take them along when visiting potential funders or others outside the center.

Input Devices

You may wish to consider the following devices for capturing and working with images:

- **Digital cameras.** The prices on digital cameras have dropped dramatically, especially for what are now considered low-end cameras. Being low-end isn't a problem, however, because most of the cost in a digital camera comes from features that improve the image resolution quality. Most display technologies, like computer screens and the web, can't even show images at the resolution of today's high-end cameras—you can see the improvement only when the images are printed by expensive printers. As a result, you can purchase perfectly fine cameras for your needs at a fraction of recent costs. If at all possible, get several cameras so that many students can work on projects simultaneously without having to wait for others to finish. Consider getting cameras that use 3.5" disks to capture and store the images. Even though many of the newer cameras use memory cards instead, disks have the advantage that they enable kids to move their photos over to the computers while the camera is being used by other students. Memory cards allow you to take more and bigger pictures, but they require cables to transfer the files to the computer, so you could create a bottleneck by tying up both the camera and the one computer it connects to. If you do get cameras that need cables, make sure that the devices are compatible with your operating system.

- **Scanners.** You may want to consider getting a flatbed scanner so that you can scan pictures and text into the computer. Scanners have become fairly inexpensive and, depending on the projects you undertake, may have certain advantages. If your budget allows for both cameras and a scanner, go for it, but if you can afford only one item (especially if you want to make photography a part of your activities), the digital camera will probably be more useful to the center than the scanner.

- **Drawing tablets.** If you're going to do serious art projects with older kids, you may want to get a pressure-sensitive drawing tablet for one or more of your computers. This device replaces a mouse with a digital pen, which feels and acts much more natural than the awkward mouse. When used with sophisticated graphics programs like Photoshop, it can even draw thick or thin lines, depending on how hard you press.

Software

During the two-year YDC Pilot, the Morino Institute sought the advice of experts and researched a number of different software applications for use in technology learning programs. They are listed in Worksheet 8.1. The following criteria were used in selecting the software applications:

- **The applications support the inquiry-based learning approach.** They allow users to produce, manipulate, publish, and exchange original visual, text, and audio content, as opposed to emphasizing the practice of rote skills or the absorption of predeveloped content.

- **They support multipurpose use.** Both beginning and advanced students can use them for simple or complex tasks.

- **They will operate over a network.** Users can import and export files across the network.

- **They are widely used.** In developing proficiency with the software applications, children and staff will learn marketable and widely applicable skills. Many of the recommended applications are commonly used by professionals worldwide.

- **They are made by established companies.** You need to be able to count on reliable customer support.

As you go about selecting software for children's learning programs, ask yourself the following questions:

- Am I overly focused on the software itself, or is my choice driven by project themes and activities?

- What kinds of skills should the students practice to meet our program goals? How will the applications introduce them to the information they need or help them produce work products related to the project?

- Does the software allow students to create original visual, text, or audio content?

- If preproduced content comes with the software, is it age-appropriate and culturally sensitive?

- Are the content and features complex and flexible enough to allow for continuous reuse and sequential skill development?

- If the content or tools are just for beginning users, does the software have a natural link to a more advanced application? What will children move on to after they learn to use the software?

- Are the graphics and sound in the software distracting, or do they enhance the experience of using the program? Can the sound and other features be adjusted or modified?

- Can the software be used for group projects (i.e., can it import or export files over a network, or does it have other collaborative features)?

Operating Systems

An operating system is the most basic piece of software on a computer. In essence, it translates what you do in an application into the language of the machine so that it can execute your commands. The Microsoft Windows operating system is the most popular operating system, especially among businesses. The other significant choice is the Macintosh OS, which runs on Apple Macintosh computers and has a following among creative professionals. The words "platform" and "environment" are sometimes used to describe the combination of hardware and operating system used on a computer.

Your operating system, in all likelihood, will be determined by the type of machines you use in your lab and may even come with the computer. Most of the software programs recommended in Worksheet 8.1 are available for all major operating systems, but unless you have no choice, try to avoid mixing platforms in your center. Having some Apples and some Windows machines is not an insurmountable obstacle, but it will put an added burden on your support team.

Your learning goals and budget will be major factors in your selection of software. In most cases you will have to buy a license for each machine that runs a particular application, and you will need enough storage capacity and memory on your hardware to run all the programs you intend to install (but not necessarily simultaneously).

Keep in mind that although the software programs listed in Worksheet 8.1 have been listed according to their main purpose, many programs, especially those for creative applications, have features that also appear in other types of programs. If you're in a budget crunch, for example, you may find that some of the outlining and charting features in a presentation program like PowerPoint will suffice to replace a thought-organization program like Inspiration. Worksheet 8.2 lists some helpful guides you can use when selecting software for children.

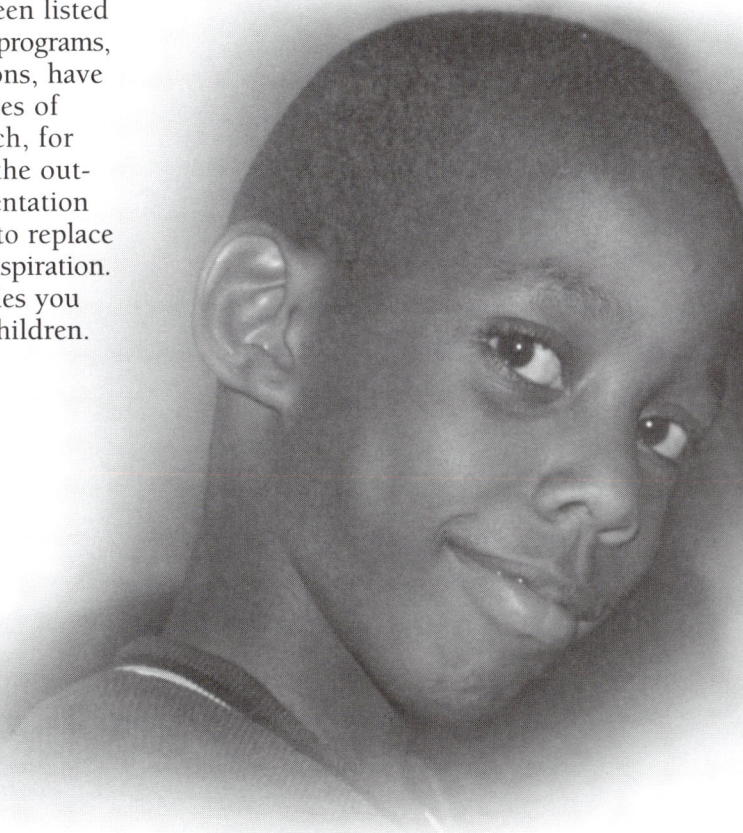

Worksheet 8.1
Checklist of Recommended Software Applications

Listed below are the basic types of software you should consider for your technology learning center. Use this list to help guide your strategy for obtaining the programs you think are most appropriate for your center.

Category	Software Title (Manufacturer)	Description	Already Own	Should Purchase/ Obtain
Organizing Thoughts and Ideas	Inspiration (Inspiration Software, Inc.)	This helpful product allows the user to generate and organize creative ideas with graphic organizers. The company also offers products for encouraging and using visual learning in the classroom. A version for young children, Kidspiration, is also available.		
Image Manipulation, Drawing, and Painting	Photoshop (Adobe Systems)	Photoshop is one of the most powerful creative programs around. It is used mainly to edit graphic images, for tasks such as making changes in photos and translating images into various file formats. It also has powerful drawing and painting tools. In many projects that incorporate digital photography, you'll have kids bring their pictures into Photoshop to adjust and enhance them, then export the pictures in a form suitable for the web or other use. Photoshop has a vast array of highly sophisticated functions that professional image editors use to prepare images for publications. Although you and your kids will probably never use those functions, by introducing kids to Photoshop you'll expose them to a real-world application that can help them get jobs.		
	KidPix Studio Deluxe (Broderbund)	KidPix is a multimedia authoring application designed specifically for children. It consists of a suite of drawing and painting tools. Its interface is child-friendly, and it omits some of the more sophisticated functions that might confuse kids and that they are unlikely to need.		
Multimedia Authoring and Presentations	HyperStudio (Roger Wagner Publishing)	HyperStudio is a multimedia authoring system that can combine words, pictures, sounds, animations, and videos into a single presentation. Its simplest form is that of the slide shows we see so often at lectures or conferences. More sophisticated features allow users to animate movement, add audio files, and create video-type effects, such as fade-outs. HyperStudio is a good, all-purpose authoring tool that is a bit more powerful than KidPix and somewhat easier for kids to use than Microsoft PowerPoint, a program that is targeted mostly at business users.		
Web Browsers	Explorer (Microsoft) Navigator(Netscape)	A web browser is the program that lets you view pages on the web. Either of these two popular programs is perfectly fine for your uses.		

(continued on next page)

Checklist of Recommended Software Applications

Category	Software Title (Manufacturer)	Description	Already Own	Should Purchase/ Obtain
Web Publishing	Netscape Communicator (Netscape) Dreamweaver (Macromedia) FrontPage (Microsoft)	Many, many choices are available in web publishing software, and new applications are on store shelves as rapidly as the Internet landscape changes. Communicator is a solid product that is both free and available on most community computers. The advantage of its widespread availability is that kids can take their files home, to libraries, or to school to continue their work. Dreamweaver and FrontPage are other good web publishing products.		
Email Clients		An email client is the software on your computer that lets you receive, read, and send email. Many types of clients and many ways to get email are available, including web-based email and free email services. Your Internet provider may supply you with clients. The web browsers Microsoft Explorer and Netscape Navigator have built-in email clients.		
Desktop Publishing	Publisher (Microsoft)			
Reference	Encarta (Microsoft)	Includes a dictionary, encyclopedia, atlas, and thesaurus		
Graphics	Painter Classic (MetaCreations)	Incorporates "natural media" features—users can paint in oils or watercolors or draw with crayons, chalk, or pastels		
Keyboarding Skills	Mavis Beacon Teaches Typing (Mindscape)			
Multimedia Authoring	Microworlds (Logo Computer Systems)	Programming environment for animation, graphics, sound, and games		
	Premiere (Adobe)	Video editing		
Basic Skills Building	Reading Magic Library Series, Graph Club, and related software (Tom Snyder Productions)	Multiple K-12 subject areas		
	Carmen Sandiego Series (Broderbund)	Geography, math, and word games		
	Sim Ant, Sim Town, Sim City (Maxis)	Ecological problem-solving games		

(continued on next page)

Worksheet 8.1 continued

Checklist of Recommended Software Applications

Category	Software Title (Manufacturer)	Description	Already Own	Should Purchase/ Obtain
Word Processing and More	Microsoft Office (Microsoft)	Microsoft Office is an integrated software package that includes a word processing application (Microsoft Word), a spreadsheet (Microsoft Excel), a multimedia authoring/presentation product (Microsoft PowerPoint), and a variety of other applications and utilities. Several editions are available (e.g., Standard and Professional), each of which provides a different combination of applications. Office provides a great deal of functionality at a good price. Moreover, its popularity in the market makes it easier to transfer files between people and gives kids exposure to tools they'll use in the real world. It is possible to buy each Office application separately and swap in some applications from other companies, but we find the package deal to be the most cost-effective and easiest to integrate. Upgrades are available at a reasonable cost to licensed owners.		

Worksheet 8.2
Checklist of Software Guides

The following list describes some of the many software guides that can help you choose software based on cost, age-appropriateness, and educational value.

Title (Creator)	Description	Notes
ESI Online www.edsoft.com (The Educational Software Institute)	Contains searchable reviews of all kinds of software, including a category for ESL software	
Multilingual Education Technology Consulting http://users.rcn.com/ abishop.interport (Ana Bishop, chairperson for instructional technology for the National Association for Bilingual Education and a multilingual education technology consultant)	Contains complete reviews of ESL software programs as well as great links to related sites	
Only the Best www.ascd.org/frameotb.html (The Association for Supervision and Curriculum Development)	Contains ASCD's annual guide, *Only the Best: The Annual Guide to the Highest-Rated Educational Software and Multimedia*	
Children's Software Reviews www.reviewcorner.com/ reviews.html (The Review Corner)	This site reviews software made for children. The ratings seem inflated (most software is "recommended" or "highly recommended"), but the reviews contain detailed information about each piece of software, including age appropriateness, skills covered, cost, and system requirements.	
Children's Software Revue www.childrenssoftware.com (Children's Software Revue)	This site offers samples of the reviews published in the print version of *Children's Software Revue*. The frank reviews detail how valuable each piece of software being reviewed really is. The reviews include target age, cost, date of the most recent evaluation, content, and usability. A link allows readers to submit their own reviews online.	
CTCNet Center Start-Up Manual: Software Selection and Criteria www.ctcnet.org/ch5.htm (CTCNet)	This chapter in CTCNet's manual provides a comprehensive overview of how to choose software for out-of-school technology programs. It covers issues such as buying pieces of software individually versus buying software "bundles" and describes the types of software needed to start a program.	

Technology Consultants

Today's nonprofit organizations exist in an environment of technology convergence. From errant fax machines to network computers, printers, and copiers, even the most knowledgeable nonprofit requires assistance with everyday technology.

To answer this call, a variety of consultants have stepped forward. All kinds of consultants are available for hire, from generic computer consultants to specific network consultants. But finding affordable, personable, knowledgeable consultants who can work with you to solve your problems and—one hopes—ward off future problems can be trying. Let's face it: Hiring an expert in a field about which you may know almost nothing is not the easiest task.

This section will familiarize you with qualities you should look for, concerns you should address, and other aspects of hiring a technical support consultant.[11]

Find Technology Consultants in Your Area

If you live in a major city, finding potential consultants is easy. You can use both local and national lists, such as Opportunity Nocs, Craig's List, Consultants OnTap, Idealist.org, and other national, regional, and local websites. Nonprofits in small towns or rural areas have to work smarter to obtain a good list of consultants.

Remember, the key is to work smarter, not harder. Don't overlook great resources such as word of mouth, foundations, circuit rider associations (discussed below), and nonprofit technical organizations.[12] Word of mouth is a common and easy way to hear about good consultants. The likelihood that you are the first agency in your region to look for technical support is small. Contact organizations of similar size to yours, and find out who their technical support providers are (or who they are not!).

Foundations are often overlooked as an information resource, but many foundations spend a significant amount of time compiling resources for their grantees. Contact one or more of your funders and ask for referrals. This step not only will keep your funders apprised of your ongoing efforts to meet the goals of your organization but also will allow them to help you in ways that do not involve grant applications.

A somewhat new but widely growing phenomenon in the nonprofit world is that of circuit riders. Circuit riders are groups of technical support consultants who split their time across multiple nonprofits to provide ongoing support. Many foundations have created circuit rider associations for their grantees.

One of the most important elements to consider in hiring a consultant is to make sure the person is a specialist in working with nonprofit organizations. Many technology consultants are accustomed to working with businesses that have large information technology budgets and that need the latest and greatest in computer technology. Make sure the person you hire is sensitive to the realities of your nonprofit.

Define Your Technical Needs

Now that you know where to start looking, you need to define your technical assistance needs. Perhaps the most valuable piece of advice regarding hiring a technical consultant is to be clear about what you want before you start the hiring process. You, the client, have a major role in ensuring that appropriate services are provided by the consultant.

The truth of most consultant-client relationships is that when one asks, "Did the client organization get what it wanted?" and "Did the client organization understand what it got?" the answer to both questions, sadly, is no.

Quite a few consultant-client relationships fail because the nonprofit never had a clear picture of what it wanted or needed in the first place. Without a clear vision of what you want, you could be sold the neighbor's barn or a mansion on Main Street and be unable to distinguish between the two.

The key to any good relationship is communication. A primary component of good communication is being able to state what it is that you want or need. You do not need to learn the entire lexicon of technical terminology, but if you want all your computers to be able to share files and folders quickly, you should be able to say, "I want to set up a system that will allow all the computers to share files and folders, and I'd like it to be fast."

Evaluate Potential Technology Consultants

It is a good idea to interview at least three consultants. You should ask the candidates to bid on the project or tasks by providing you with a formal written proposal for the work. Project proposals give you many advantages:

- Project rates are generally cheaper than hourly rates.

- Reviewing the proposal gives you time to research components of the proposed work until you have a general understanding of what is being offered. Unfortunately, interviews don't offer that luxury.

- Proposals give you a glimpse of the working process of your consultant. Proposals that are well written and well organized reflect consultants who have pride in their work.

- A proposal allows you to see in writing whether the consultant can translate technical information into a common language that you and your staff can understand.

Make the Hiring Decision

Consultants who provide expert advice, hands-on technical assistance, and a willingness to apprentice staff members and volunteers when they run into difficulty provide the best support. Ultimately, getting expert technical support should accomplish two goals: solving the problem at hand and teaching the user to solve the same problem should it occur in the future. A consultant who does everything for you or who works only after hours will not be able to work with your organization to accomplish the second part of the support goal.

When you are choosing among consultants, communication skills are often a determining factor:

- Did the consultant seem to understand you?

- Did you understand the consultant?

- Was the consultant patient?

- Did he or she listen to you and understand what your organization does?

- Will he or she fit in with your organizational working style?

- How available will the consultant be for providing assistance?

Don't Settle for Mediocrity

After you have made your choice, keep the consultant-client relationship open and honest. If you do not like the level of service you are receiving, discuss the matter with the consultant. If you do like the service and quality of work, discuss that as well. Feedback for the consultant allows him or her to gauge better how the work is being perceived.

If the relationship with the consultant is unproductive, end it. Many organizations spend years in unproductive client-consultant relationships before seeking other assistance. Whether you have a three-month, six-month, year-long, or multiyear project for a consultant, establish an evaluation period. Ninety days is generally more than enough time to get a sense of how well your staff and the consultant work together. This evaluation period benefits the consultant as well. Scheduling a "time out" at the beginning of the contract allows you both to step back and evaluate the process before continuing.

Youth Safety Online

Most people, including youth, have a fun, safe trip on the information superhighway. Although most online experiences are positive, cyberspace does have a dark side: It includes people who attempt to exploit children and others through the Internet as well as materials on the Internet that are adult-oriented and inappropriate for children.[13]

Fear of exploitation and abuse or fear of exposure to inappropriate material shouldn't prevent a school, youth group, community-based organization, or parents from allowing young people to use the Internet. An organization can use various simple measures to ensure the safety of youth online.

Although some highly publicized cases of abuse involving computers have occurred, reported cases of harassment and abuse because of a child's online activities are infrequent. Of course, like most crimes against children, many cases go unreported, especially if the child is engaged in an activity that he or she does not want to discuss with a parent. *Child Safety on the Information Highway*, by the National Center for Missing and Exploited Children, notes that "the fact that crimes are being committed online . . . is not a reason to avoid using these services." To tell children to stop using online services would be like telling them to forgo attending college because students are sometimes victimized on campus. A better strategy would be for children to learn how to be "street smart" in order to better safeguard themselves in any potentially dangerous situation.

Online Safety Guidelines

A fear-based approach to online safety guidelines is not advisable for organizations; a culture of fear can lead to so much distrust that it defeats the purpose and benefits of the Internet and positive online interactivity. Any program has risks, whether online or face-to-face. Exercising common sense, adapting your existing offline prevention systems to cyberspace, following the law, educating participants, establishing good tracking of children's online activities, and supervising online interactions are the best online safety measures.

The most effective way to prevent youth from using the Internet for inappropriate activities is to teach them how to use the Internet and related technologies within the context of well-organized, purposeful, and engaging activities in an adult-supervised environment. In other words, if children learn how to use the Internet and multimedia technologies in ways that are positive, constructive, and meaningful, they will have considerably less interest in—and opportunity for—using the Internet for negative or meaningless activity.

Children's Privacy

As a result of the Children's Online Privacy Protection Act (COPPA), which Congress passed in 1998, the Federal Trade Commission (FTC) adopted new rules on how privacy policies should be posted and what companies need to do to comply with the new prohibition on collecting personal information from young people without parents' permission. The rules took effect in April 2000 and require that all websites that gather information from children under age 13 first gain "verifiable parental consent."[14] For computer use in schools, the rules allow teachers to act as parents' agents or intermediaries.

The FTC allows websites to vary how they gain permission, depending on what information is being gathered and how it will be used. For example, websites are required to use reliable forms of consent, such as postal mail, fax, credit card, or "digital signatures," before children can participate in chat rooms or give out personal information that will be made available to third parties. If the site is using the information only internally, however, the operators can accept email from parents, as long as a follow-up email or phone call is made to them.

@ *For more information, examples, and resources, see the extensive YouthLearn website at www.youthlearn.org*

chapter 9

PUT IT ALL TOGETHER:
INQUIRY-BASED PROJECTS IN ACTION

Overview

The projects in this chapter were created to help you develop your own projects that use and teach the tools of technology. Don't worry about not being able to do everything exactly as written; there is a lot of detail, and only a very advanced teacher or youth worker will be able to master every session the first time around. Every group of kids is different; use the sample projects as a guide, but do not be discouraged if they don't quite fit your group or if you have to cut out parts as you go. Use what is helpful to you and make notes about what works and what doesn't.

In many sections we have included links to sites that we think are helpful in advancing the project. The URLs (website addresses) we provide here were accurate at the time of publication, but the dynamic nature of the web means that they may have changed in the meantime. Don't give up on a site too quickly; a little digging and searching on your part may turn up the old site at a new address. Of course, you can also look for sites on your own and tailor the projects to your group's particular needs and interests.

How to Use the Projects

The Soil Around Us and *Civil Rights Through a Lens* are inquiry-based projects that will help your group of young people formulate questions about things they are curious about. Your job during the course of the project will be to help group members find ways to investigate and answer their questions and to document, analyze, and present their findings. Throughout the project, they will practice:

- Cognitive skills—reading, writing, researching, interviewing, measuring, and labeling

- Collaborative reading, writing, and groupwork processes—brainstorming, labeling, mapping, interviewing, and storyboarding

- Social skills—working in pairs and groups, sharing, listening, taking turns, and assisting others

- Computer-based activities emphasizing multimedia skills—creating and editing drawings, photos, and text, developing basic Internet navigation skills, using multimedia authoring software, and presenting information in multimedia formats.

The projects in this chapter can be adapted to a variety of interests and ways of working. Each can be done in sequence or out of sequence, in parts or as a continuous, connected project. Because many out-of-school programs follow an academic calendar of winter, spring, and summer terms, we suggest you plan an 8-week project.

Contents
- General advice for implementing inquiry-based projects
- *The Soil Around Us*, a sample project for 9- to 11-year-olds
- *Civil Rights Through a Lens*, a sample project for 12- to 14-year-olds

Audience
- Youth development staff
- Teachers and center instructors

You can find more information and links to other resources on the topics discussed in this chapter by visiting the YouthLearn website at *www.youthlearn.org*

Assuming a typical term of 12 to 13 weeks, this schedule allows extra time for field trips, preparatory activities such as group introductions, closure activities such as family nights, and extension activities.

Things to Do in Every Session

Certain activities are important enough that you should do them in every session with your group. Especially important are community builders, reading (both aloud and silently), and journal activities.

Community builders are short activities that help break up the day. They can be used to help kids feel comfortable with one another, to ease the transition between longer activities, or to help instructors gain control or focus in a class that is starting to stray or act out.

Reading is a critical component and should be done both as a group and individually. Kids can feel awkward reading aloud in a group at first, but it will help them develop listening skills and energize them about the subject you are studying. Reading silently can be a great end-of-the-day activity and will help reinforce the importance of reading. Remember that when the kids are reading individually, you should too. For more about teaching reading skills, see page 94.

Journal activities cover a wide range of possibilities, which could include recording notes, writing creatively, and responding to questions as well as drawing pictures and creating collages. A routine of using journals will help kids form a habit of writing for their own personal reflection and enjoyment. You can use journals to pose questions that connect to reading materials and project activities. Remember to be creative and build your own journal along with the group. For more about how to incorporate journals into your daily sessions, see page 99.

Examples of these activities are given throughout the projects—particularly in the early sessions—to give you an idea of what they look like and how they fit into an individual session. As you get to know your group, feel free to mix up these activities, move them around, and use your creativity to invent new ones.

And again, don't hesitate to go to www.youth learn.org for more information and online resources.

Guidelines for Your Group

Age

The young people participating in either project should be organized into groups with no more than a three-year age span. Because levels of skill and maturity vary, use your best judgment when placing them in groups. A reference book on teaching and child development will help you make decisions about how to customize the project activities for your groups. Check out *Yardsticks: Children in the Classroom Ages 4-14; A Resource for Parents and Teachers*, by Chip Wood (Northeast Foundation for Children, 1997), or a similar resource.

Organize the young people into groups with no fewer than 8 and no more than 20 children for each adult. Ratios should be determined according to your comfort level and the needs and comfort level of the group.

Time

A good child development book like *Yardsticks* will help you decide how long each session should be. For 9- to 11-year-olds, plan to spend at least 60 minutes per session; this should allow you enough time to accomplish the basic components of the session. This age group can handle longer sessions (two hours or more) if the activities are broken up with snacks and rest periods. For 12- to 14-year-olds, scheduling longer sessions (90 minutes or more) generally works better, even if it means having the participants meet for fewer sessions each week (for example, one long session per week versus two short sessions). The longer sessions allow time for discussions and independent work, which teens generally prefer. For both groups, remember to schedule time for group welcome and closure activities during every session.

Space

The group will need a comfortable space for project activities. It should have at least one large table and several chairs for writing, drawing, and other sit-down activities. (An even better scenario is to have several worktables that can be placed in different arrangements for small group activities.) You will also need a secure space to store project supplies and a place to hang materials such as drawings and maps.

Materials

Specific supplies needed for each project activity are listed in the sections describing the sample project sessions. In general, it will be helpful to have the following items on hand:

- Scissors (child and adult size)
- Clipboards
- Glue sticks
- Post-it pads
- Masking tape
- Scotch tape
- Index cards
- Loose white copy paper
- Colored construction paper
- Thin felt-tip colored markers
- Thick colored markers
- Pencils
- Pens
- Crayons
- A single-hole punch
- A ball of string
- Rulers
- Manila folders
- Posterboard
- Newsprint pads, flip-chart paper, or a roll of butcher paper
- Composition books or spiral notebooks (composition books are preferable)
- A children's dictionary
- A standard adult dictionary
- A thesaurus
- A map of the world, globe, or atlas

Kids should be encouraged to use online dictionaries, encyclopedias, and map sites. It is also helpful to have a cassette recorder to record sounds and play tapes. A hand-held recorder can be used during field trips and in the classroom to record interviews, music, or interesting sounds.

Technical Needs

Keep in mind that you don't have to have all the equipment listed in this section to do all the activities in each project. Be creative—if you don't have an Internet connection or can't get a specific kind of software, substitute other activ-

ities. (But don't give up too easily; there are many resources out there for donated software and equipment—see the Technologies section in www.youthlearn.org for a list of organizations that specialize in donated computer goods for nonprofits.) If you do have the resources, however, here are our recommendations for the hardware, the software, and the Internet connection you should have for these projects.

Hardware and Accessories

- **CPUs.** A PC of at least Pentium III speed or an equivalent Macintosh (an iMac or Power Mac).
- **Sound cards.** It is helpful to have computers that are outfitted with sound cards (most computers that can run multimedia software and the Internet have these cards).
- **Microphones.** It is also helpful to have at least one microphone that can be plugged into a computer to record sound.
- **Digital cameras.** For prices and recommendations on digital cameras, check vendor sites or go to www.cnet.com. The Sony Digital Mavica MVC FD81 series camera is an excellent choice because it stores pictures on removable floppy disks. Cameras with floppy-disk recording capabilities are more expensive but easier to use than cameras that must be hooked up to a computer to download the pictures. The Sony Digital Mavica FD81 can also record short video clips (up to 60 seconds) with sound. If a digital camera is not available, a Polaroid camera (classic or iZone) makes a good substitute.

These lessons do not require one computer per child. In fact, for many activities we recommend that each child share a computer with a partner to benefit from peer coaching. Kids can rotate between computer and noncomputer activities during a project session.

Activities that require access to the Internet (email or the web) can be done with just one Internet connection, if necessary.

Software

We have included software suggestions within the instructions for each session. We recommend the following types of software programs:

- **Drawing.** A program that young children can use to draw freehand and to manipulate clip art (KidPix or a similar application).

- **Multimedia.** A program that children can use to make multimedia presentations (KidPix, HyperStudio, PowerPoint, or a similar application).

- **Photo editing.** A program that children can use to edit digital photos (Adobe Photoshop, Adobe Photoshop Elements, Microsoft PhotoEditor, or a similar application).

- **Brainstorming.** A program that kids can use for mapping and webbing exercises (Inspiration for grades 4-12, Kidspiration for grades K-3).

- **Word processing.** A program for writing and manipulating text that also can import drawings and photos (Microsoft Word, Broderbund Print Shop, or a similar application).

- **Graphing/chartmaking.** A program that young children can use to make graphs and charts. (The Graph Club by Tom Snyder Productions—www.teachtsp.com— is specifically designed for children from kindergarten through fourth grade.) A business application that makes graphs and charts, such as Microsoft Excel or ClarisWorks, could also be used for older kids.

- **Web publishing.** A WYSIWYG (what you see is what you get) program for creating web pages (Netscape Composer [a component of Navigator and Communicator], Dreamweaver, or a similar application).

See page 110 for more information about choosing hardware and software.

Remember, the focus of these projects is on learning and youth development. Don't get hung up on the equipment. The technology is merely a means to the goal of providing kids with creative and exciting learning experiences.

Check out StargazerNET (www.stargazernet.net) for a broad range of online tools.

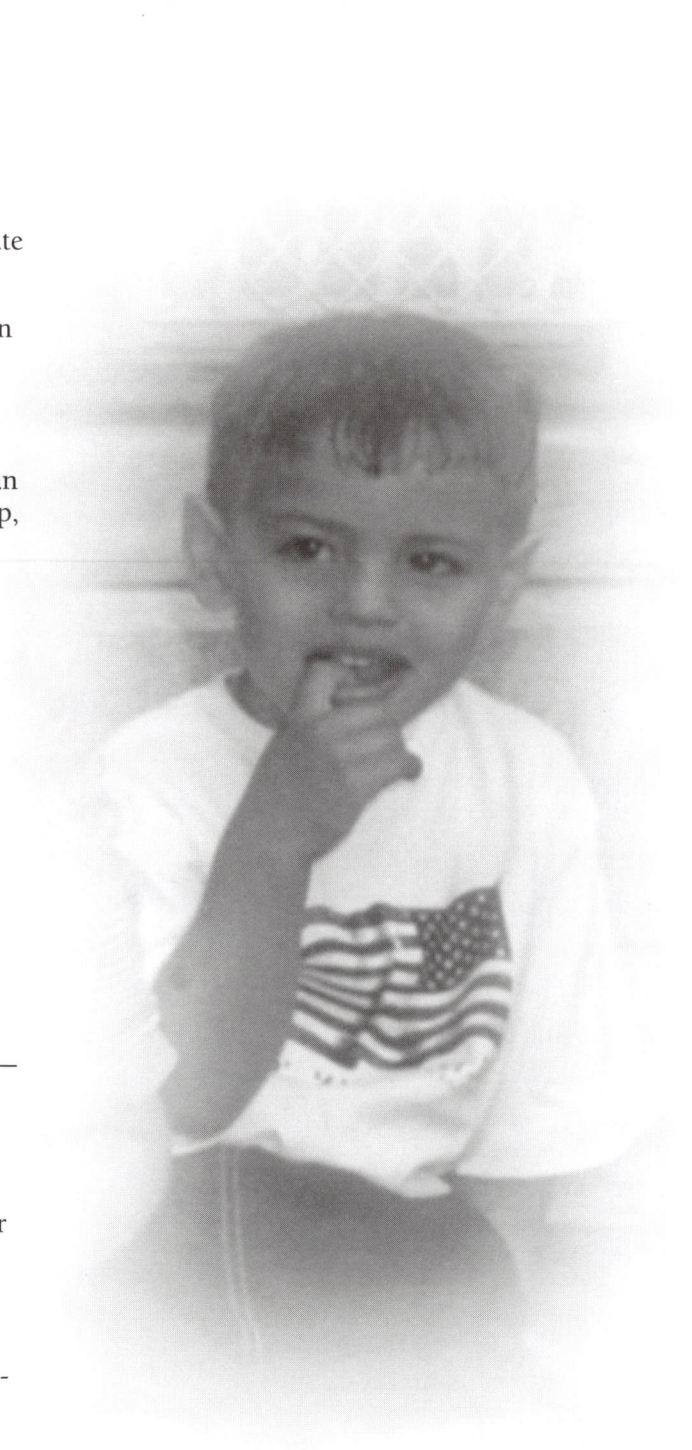

Notes

★ The Soil Around Us (for ages 9 to 11)

Goals of the Project

The Soil Around Us introduces 9- to 11-year-olds to methods for investigating the origins, characteristics, and uses of soil. This inquiry-based project builds on kids' innate curiosity about the natural world. The project is designed for this age group because the environment, environmental protection, animal and plant life, natural systems, and life cycles are frequent themes in their schoolwork and in their out-of-school lives.

The project can be adapted for other age groups by modifying reading material, group activities, websites, and software. The project is designed for eight sessions, each about 90 minutes in length. Project activities combine outdoor education, indoor hands-on learning, and the Internet and related technologies.

Components of the Project

- **Reading**—you will read to the group and have group members read on their own during every session.

- **Writing**—students will practice writing during every session, in both individual and group journals.

- **Oral presentation**—each student will practice speaking in front of the group and listening to others speak.

- **Presentation of quantitative data**—the group will collect, analyze, and organize numeric information during one or more sessions.

- **Organization of information**—the group will sort and categorize information on maps and charts.

- **Visual communication**—the group will take and edit photos, and draw by hand and with software tools.

Questions to Investigate

Helping the group identify questions they want to investigate is the critical first step of any inquiry-based project. To guide this process, it is helpful for you to pose your own questions prior to starting the project.

Your questions could be broad, such as:

- Where can we find dirt?

- What is dirt made of?

- Where does dirt come from?

- What things live in dirt?

- Are there different kinds of dirt?

- What things need dirt to grow?

or questions could be more specific to a particular topic, such as:

- Why do worms come out of the dirt when it rains?

- What do ants do in an anthill?

- What happens in winter to moles and other animals that live underground?

- What can we use dirt for?

- Is dirt different in different places in the world?

(continued on next page)

Books for the Project

In addition to the standard materials and equipment suggested in this chapter's overview, we recommend the following children's books (books with similar themes can be added or substituted):

- *Filet of Soil,* by Barry Rudner (Windword Press, 1996). A rhyming story about the characteristics and virtues of soil, including a glossary.

- *A Handful of Dirt,* by Raymond Bial (Walker & Co. Library, 2000). Explains how soil is created through natural decomposition processes and describes the living organisms in soil. Features great color photographs.

- *Sand,* by Ellen Prager (National Geographic Society, 2000). Explains how different types of sand are created, where sand can be found, and how it gets there. Includes drawings, color photographs, and instructions on how to make sand.

- *One Small Square Backyard,* by Donald Silver (McGraw-Hill Professional Publishing, 1997). A guidebook on outdoor education activities that can be conducted in a backyard, including suggestions for simple science exploration activities.

- *Our Endangered Planet: Soil,* by Suzanne Winckler (Lerner Publications Company, 1993). Discusses how different types of soil have been endangered through farming, mining, and grazing. Includes a glossary and color photographs.

9

Session 1
Brainstorming

Goals for the Session	• Create KWHL chart (see Activity 3 for details). • Introduce journals.
Preparation	• Prepare for the read-aloud of *Filet of Soil* by reading the text, looking at the pictures, and thinking about questions for discussion.
Materials and Equipment	• A copy of *Filet of Soil*, by Barry Rudner • Digital camera(s) • Newsprint or flip-chart paper • Colored markers • Post-it pads • Composition books (one for each group member, including instructors) • Pens and pencils • Construction paper • Glue sticks • Crayons • Magazines

Activity 1: Community Builder

Sample Community Builder: Circle Mirror

Students and instructors stand in a circle, allowing room between people for arm movement. You should initially take the lead and play the role of the "mirror." Lead the group in a motion (e.g., moving one arm like a windshield wiper, marching in place, or another whole-body movement) for approximately 30 seconds, then say "Freeze!"—whereupon all the "reflections" stop moving and hold their positions. Select a person to take your place, and let the reflections begin to imitate the motions of the new mirror. One child can do the timekeeping.

Activity 2: Read-Aloud

Reading aloud is another important activity to include in every session. The suggested book for this session is *Filet of Soil*, by Barry Rudner.

During the read-aloud, show the pictures in the book to the group. After reading, ask the group to discuss a few questions. Have the kids pair up to discuss the questions with a partner, and then have them report back to the whole group.

Sample Questions

• Make a list of all the different words the book uses to describe soil. Can you think of other words?

• The book points out that the way we sometimes use the word "dirty" (as in "dirty look") shows that soil has a "bad reputation." Why do you think we associate dirt and dirtiness with bad things? What are the good things about dirt?

• How would you describe soil to someone who does not know what it is?

You may also want to encourage children to bring in any books they have at home that relate to the topic of soil or sand.

Activity 3: Create a KWHL Chart

A KWHL chart is a brainstorming technique that visually displays the information kids already know and want to know about a topic. The letters K-W-H-L represent four columns of information:

- "What do we KNOW?"

- "What do we WANT to know?"

- "HOW can we find out what we want to learn?"

- "What have we LEARNED?"

The last column is usually filled in after the group has participated in several investigative exercises.

A sample KWHL chart is shown below.

Label the four columns on a sheet of newsprint or flip-chart paper before working with the group, as shown below. Have the group members sit on the floor in front of the chart and brainstorm answers to the questions. Lead the brainstorming session and offer one or two examples for each column to get the group moving in the right direction.

The chart should remain posted in the classroom throughout the sessions. Children should feel free to add or delete information from the chart as they participate in further activities. A group member's answer does not have to be correct to be included in the chart. As they participate in the project, the kids will draw their own conclusions and correct their own assumptions.

Activity 4: Journal Activity

For the first journal exercise, ask the group to respond to a question that requires creative problem solving and imagination. You can use something related to your project, such as "In words and pictures, describe how we can solve the problem of pollution." Or you can choose something completely different to get the creative juices flowing, like "How would you design a rocket ship for traveling to Mars?"

Journal Exercise

1. Post the question on a board or piece of paper where everyone can see it. You should also respond to the question in your own journal.

2. First, you share your response in front of the group (i.e., show your journal, explain the words and pictures used, and then pass it around). Next, ask at least one participant to share his or her journal entry in front of the whole group.

3. You can then ask two children to volunteer to share their journal entries with each other in front of the group. Following the demonstration, have all members of the group share their entries in pairs. (For more about the pair-share technique, see page 91.)

4. Finally, all students can share their responses with the whole group.

After they have recorded and shared their

Sample KWHL Chart

K	W	H	L
What do we KNOW?	**What do we WANT to know?**	**HOW can we find out what we want to learn?**	**What have we LEARNED?**
• Dirt is dark brown. • My mother won't allow dirt in my house.	• What is dirt made of? • What things live in dirt?	• We could look at dirt outside. • We could find information on the Internet.	• Not all soil is the same. • Organisms live in soil.

9

entries, have the kids decorate their books with construction paper, magazine clippings, markers, and so on. Decorating provides an opportunity for the children to take full ownership of their journals as a space for personal expression. Allow at least 20 minutes for decoration.

Tips

- Keep your own journal and make entries every time the group is asked to write. This activity presents a model and allows you to participate in group sharing of entries.

- Ballpoint or felt-tip pens (preferably in a variety of colors) are a good choice for journal writing because they cannot be erased. When erasing is an option, children spend too much time deleting their work and not enough time moving forward.

Children who have not received instruction in drawing often become easily frustrated by drawing tasks, both on paper and on a computer. To help a group become more confident with drawing, try setting aside time in each session to work on exercises from the book *Drawing with Children: A Creative Method for Adult Beginners, Too*, by Mona Brookes (J. P. Tarcher, 1996). The exercises are easy to follow, even for instructors who have had little experience with drawing. Try doing one new exercise per session.

Activity 5: Optional Extension Activities

- Start a glossary for the project on index cards or large sheets of paper. Post them on a wall. The definitions in *Filet of Soil* can be used to get the ball rolling. Add new words as they come up in the course of readings and activities.

- Collect samples of the different kinds of soil described in *Filet of Soil*: dirt, mud, dust, soot, and so on. If necessary, mix soil with water or leave it out to dry to match the consistency given in the descriptions. Label the samples.

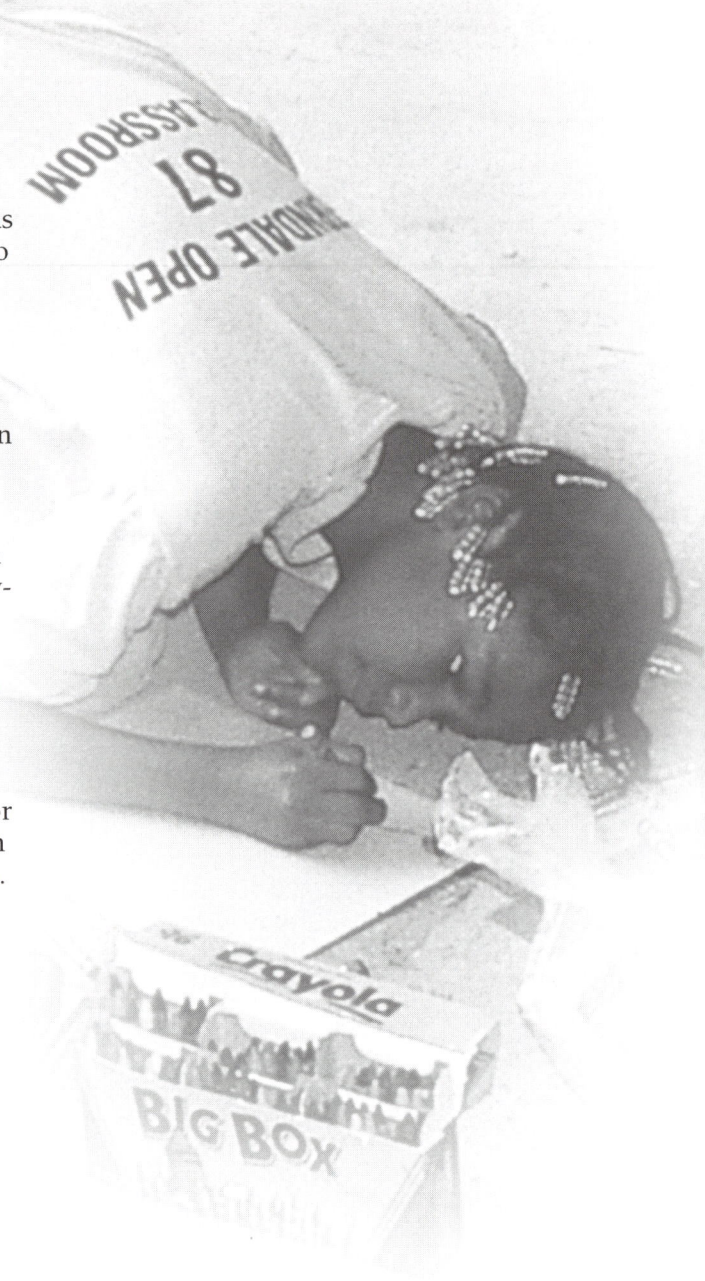

Session 2
Mapping

Goals for the Session	• Complete a mapping exercise. • Complete a webbing exercise. • Introduce the group to Inspiration software (use Kidspiration for younger children).
Preparation	• Read *A Handful of Dirt* in preparation for the read-aloud. • Familiarize yourself with Inspiration or other brainstorming software if you plan to use it during the session.
Materials and Equipment	• A copy of *A Handful of Dirt,* by Raymond Bial • Software to make diagrams (Inspiration or a similar application) • Newsprint or flip-chart paper • Colored markers • Composition books • Pens

Activity 1: Community Builder

Sample Community Builder: Zoom

Sit down with the group on the floor in a circle. Start the game by turning to the child on one side of you, looking him or her in the face, and saying "Zoom!" That child then turns to the next child and says "Zoom." This continues until the action comes back to you, at which time you can continue the game in the opposite direction. The object of the game is to have the children focus on being ready for their turn to say "Zoom" and keep the rhythm going all around the circle.

Activity 2: Read-Aloud

The suggested reading material is *A Handful of Dirt* by Raymond Bial. Have kids pair up to discuss questions, and then share the answers as a group.

Suggested Questions

• What is soil made of?

• The book says, "Without soil there would be no life on Earth." What are the reasons for this?

• Imagine that dirt no longer existed. What would life be like?

• As the book notes, "Many people think of soil as unclean." In what ways do the bacteria and other living organisms in soil help keep people healthy?

Activity 3: Mapping

Mapping is a technique for generating, sharing, and displaying ideas. In a mapping exercise, small groups of kids create hand-drawn maps made up of concentric circles. Kids at all reading and writing levels can contribute. Drawings, single words, and word phrases can all be used to represent ideas.

The purpose of the map in this lesson is to brainstorm ideas for field trip activities for the project. Each group should have several pieces of newsprint or butcher paper and a set of markers. To begin, write a word, such as "dirt," in the center of a sheet of paper and draw a circle around the word. Ask a question that relates to the word, and have the group respond by calling individual students up to write their answers (or draw pictures, if appropriate) around the center circle. Add another circle that encompasses the first to hold responses to each additional question.

If the group is small (10 or fewer children), do the first map as one group. As the children gain more experience with mapping, they may prefer to break into smaller groups to work independently. If the group has more than 10 people, break it into small groups of four to six to work on separate maps.

Suggested Map Ideas
Central word: Dirt

- Question for the first circle: Where can we find dirt in the neighborhood?

- Question for the second circle: What other places can we visit to find dirt and learn about dirt?

- Question for the third circle: What people can help us learn more about dirt?

- Question for the fourth circle: What things can we do to learn more about dirt?

Activity 4: Webbing

Webbing is another way to help kids organize and make connections between words and their ideas.

Start by demonstrating how to make a basic web map. On a piece of paper large enough for the whole group to see, make a diagram of easy-to-list information, such as the characteristics of someone in the room. Write the person's name in the center of the paper and circle it. Quickly write characteristics of that person around it. Circle each of those characteristics and connect it with a line to the center circle. Continue by branching off of each of the new words with more descriptive words. For example, for a student named Kate, you might have descriptive phrases such as "likes sports" and "friendly," and then you might expand on "likes sports" by connecting "soccer" and "ice skating" to it, and so on.

Next, ask two of the kids to come before the group and make similar diagrams by charting each other's characteristics. Then ask the group to make a diagram related to *A Handful of Dirt*. Have them work in pairs. Partners may add drawings or clip art to their diagrams.

Suggested Diagrams
- Diagram the food chain of the living things that depend on soil. Start with the smallest being and move to the largest. What do the microorganisms that live in soil eat? What eats the microorganisms? And so on...

- Diagram the process by which something that was once alive, such as a plant or an animal, turns into dirt after it dies.

If you have brainstorming software (e.g., Inspiration, Kidspiration), demonstrate to the group how to use computers to create web maps. Inspiration is an easy-to-use application for making diagrams, outlines, word webs, flow charts, and other graphical representations that organize information. Photos and other images can be imported into Inspiration to enhance diagrams, charts, and webs. (Note: Microsoft Word has box, circle, and line tools that can also be used to make charts and diagrams. Inspiration is easier to use, however, and offers enhanced functions.)

Children will not be able to remember long demonstrations, so introduce new software by showing only a few tools and functions at a time and by giving simple, focused assignments as a basis for practical exploration. They will discover many special tools and functions on their own as they use the software.

Activity 5: Journal Activity

Suggested Exercises
- In words and pictures, create a profile of an organism that lives in soil. What does it look like? How does it move? What does it like to do? What does it eat?

- Imagine that you are going to grow the best garden ever. You need a compost heap to make nutritious soil for the garden. What would you put in it? How would you build it? What other things would be in the garden?

Activity 6: Optional Extension Activities
- Start making your own soil with a compost bin similar to the one described in *A Handful of Dirt*. Check out "The Yuckiest Site on the Internet" (yucky.kids.discovery.com/flash/worm/pg000224.html) for instructions on making a worm composting bin. You will need worms, compostable trash (e.g., banana peels and egg shells), a few pieces of wood, and lining materials (e.g., newspapers or wet leaves).

- Use a microscope or magnifying glass to look closely at moist dirt. Make a list or a chart of the things you see.

Session 3
Internet Research

Goals for the Session	• Create a sand/soil compare-and-contrast chart. • Conduct website research.
Preparation	• Prepare an email or other document containing hyperlinks to the websites you plan to use during the lesson. It is easier for kids to access the websites if they can work from one document that has the links for the sites already embedded. A closed page of links will also reduce the likelihood that the students will come across sites containing inappropriate content. You can use a web publishing program such as Netscape Composer to create a simple page containing links to the appropriate sites. This page can be placed on the computer desktops for easy access. • Prepare a list of search questions to be answered during the scavenger hunt or website evaluation activity in Part 3 of this session. • Read *Sand*, by Ellen Prager, in preparation for the read-aloud.
Materials and Equipment	• A computer with Internet and email access • A copy of *Sand*, by Ellen Prager • A cup of dirt • A cup of sand • A magnifying glass • Software to make diagrams (Inspiration or a similar application) • Copies of website review sheets • Newsprint or flip-chart paper • Colored markers • Composition books • Clipboards • Markers • Pencils and pens

Activity 1: Community Builder

Sample Community Builder: Name Game

Have the group sit in a circle on the floor. The first child says her first name and an animal that she thinks represents who she is. The next child repeats what the first child said and adds his or her own name and type of animal. Continue around the circle with each child repeating all the previous children's names and animals and adding on his or her own name at the end. (Everyone should try to think of a different animal.)

Activity 2: Read-Aloud

Suggested reading is *Sand*, by Ellen Prager. Have kids pair up to discuss questions, and then share the answers as a group.

Sample Questions

• Was it surprising to learn that "sand" is the word used to describe the size of a grain? What is the difference between sand and gravel?

• What colors of sand have you seen at the beach or in a sandbox? Has anyone ever seen sand with an unusual color, such as black or pink? What do you think the sand you have seen was made of?

Tip

If a sample of sand and a magnifying glass are available, this is a great time to have the group touch the sand and look at it with the glass.

Activity 3: Scavenger Hunt/ Website Exploration

In this exercise, students look at websites containing information about soil and related subjects such as earthworms.

Best Option: Create a Scavenger Hunt

Explore the sites carefully yourself to find interesting information. From that information, develop a list of questions for the group to answer as they look at the sites. Be sure to make an answer key for your own use.

Another Option: Site Evaluation

Create a website evaluation sheet to help kids analyze the sites they visit. Be sure the sheet includes space for the students to write down at least five interesting facts learned from the sites. One example is Kathy Schrock's *Guide for Educators* (www.school.discovery.com/schrock guide/evalelem.html).

1. Gather the group around one computer and pull up the websites to demonstrate what to look for and how to navigate the pages. If the group is new to using web browsers, emphasize basic navigation techniques, such as using the navigation buttons, scrolling, and clicking on links.

2. Have kids pair up to work on their scavenger hunts or review sheets. After a period of no longer than 20 minutes, have each pair share the notes they have taken and any interesting photos, graphics, charts, or other information they have found.

Suggested Sites

- **NASA Soil Science Education Home Page** (http://ltpwww.gsfc.nasa.gov/globe/index. htm). Although it is not especially appealing visually, this site has plenty of useful, interesting information, including a soil photo gallery and answers to questions such as "How much soil is there on Earth?" The site includes links to other soil-related sites and a book list.

- **U.S. Department of Agriculture: Ask the Answer Worm** (www.nhq.nrcs.usda.gov/CCS/ squirm/skworm.html). Answers to 13 basic questions about soil. The site is easy to read and understand and includes illustrations.

- **U.S. Environmental Protection Agency, Gulf of Mexico Program: What on Earth Is Soil?** (pelican.gmpo.gov/edresources/ soil.html). A fact sheet on soil from the U.S. Department of Agriculture.

- **Yuckiest Site on the Internet: Worm World** (www.yucky.com/noflash/worm/). An attractive science education site with lots of graphics and interesting info. Check out the earthworm section to see videos of an earthworm's heart beating and a worm emerging from a cocoon.

Tip

Thoroughly preview all sites before directing kids to them. Looking over the sites will allow you to point out particularly interesting areas and will ensure that kids are not directed to pages with inappropriate content.

Activity 4: Journal Activity

Suggested Exercise

Create a compare-and-contrast chart about soil and sand. Divide a page into two columns, one for soil and one for sand. What is each substance made of? What does each look like? What lives in each one? Where is each found? How does each help the environment? How does each help animals and plant life? How does each help people? (If soil and sand samples are available, have the group take notes after touching and looking at the samples.)

Activity 5: Optional Extension Activities

- Do the sand-making activity described in the back pages of *Sand*, by Ellen Prager. Complete instructions are provided. You will need an empty coffee can with a lid, water, small rocks, a plastic cup, and a magnifying glass.

- Examine the bodies of earthworms to learn more about them. For activity suggestions, check out a related lesson plan on the San Francisco Exploratorium website (www.exp loratorium.edu/IFI/resources/doesaworm have.html). You will need worms, paper towels, a water spray bottle, alcohol, vanilla extract, sand, soil, a flashlight, a plastic tray, ice cubes, and a magnifying glass.

9

Session 4
Field Trip Preparation

Goals for the Session	• Look at books and web materials related to your upcoming field trip. • Plan/prepare roles and the worksheets for the field trip.

Preparation	Prior to this session, take a look at the Chicago Field Museum's Underground Network site (www.fieldmuseum.org/ua/nettop.htm). The site includes three soil experiment activities, one each on soil structure, scientific variables, and organisms. Be sure to read the teacher's guide for suggestions on how to introduce and manage each activity.
	
	It is strongly suggested that your field trip (see Session 5) be aligned with the activities described in the Underground Network project. This "field trip" can be conducted in a patch of healthy grass or soil anywhere you can find it—right outside your building, in a backyard, or at a neighborhood park.
	
	To navigate through the Underground Network site, first click on and read "field site" and then "journal." Then you'll have access to the soil experiment activities, which are divided into the following three sections:
	
	1. Soil structure
	    • Perform a compaction test.
	    • Perform a percolation test.
	    • Perform a texture test.
	    • Make a mudshake.
	
	2. Scientific variables
	    • Measure air temperature.
	    • Measure soil temperature.
	    • Measure cloud cover.
	
	3. Organisms
	    • Identify, count, draw, and write about insects and worms found in the soil.
	
	Make copies of the Underground Network's field trip journal worksheets (www.fieldmuseum.org/ua/images/journal.pdf).
	
	The soil experiment activities can be done in one session or broken up into two or three sessions. If your group has ideas for other field trips (e.g., a visit to a natural history museum, park, beach, or community garden), that's terrific! For general suggestions on preparing for a field trip, check out the article on field trips at www.youthlearn.org.

Materials and Equipment	• A computer with Internet and email access • Digital camera(s) • A copy of *One Small Square Backyard,* by Donald Silver • Copies of the Underground Network's field trip journal worksheets • Colored markers • Composition books • Pens

9

Activity 1: Community Builder

Sample Community Builder: Fruit Basket

The group sits in a circle on chairs (with no extra chairs in the circle), except for one person who stands in the middle. The group members sitting in chairs are equally divided among three fruits (e.g. apples, oranges and pineapples). The middle person then calls out the name of one of the fruits (e.g. apples), and all the apples plus the person in the middle change seats (who then becomes an apple, or whichever fruit he or she called). The person left without a seat becomes the next caller. If a caller says "fruit basket" all players must change seats.

Activity 2: Book Introduction

(in place of a read-aloud)

Use this time to introduce books that can be used as reference books for the field trip and the follow-up activities. Donald Silver's *One Small Square Backyard* is highly recommended.

One Small Square Backyard contains directions for outdoor education activities that can be conducted in a typical backyard. It features great illustrations and detailed descriptions of the insects, plants, and animals that can be found there. It also includes suggestions for simple science exploration activities, such as making leaf rubbings and growing mold.

One Small Square Backyard is not good for reading aloud, however. It was written as a guidebook providing instructions for outdoor education activities and is very detailed. Give a quick overview (spending no more than two or three minutes) of the book to the group. Show them some of the pages, highlight particularly interesting parts, and pass it around.

Activity 3: Field Trip Journal Worksheet Review

Pass out a complete set of the Underground Network Field Trip Journal Worksheets (www.fieldmuseum.org/ua/images/journal.pdf) to each participant. Give an overview of the sheets and explain how they will be used.

It is helpful to have the soil experiment activity descriptions (www.fieldmuseum.org/ua/net soil.htm) printed out as a reference.

Activity 4: Web Exploration

Have the kids explore the Chicago Field Museum Underground Network website to build excitement and collect information about the field trip, where they will conduct soil experiments. Children can review the website individually or in pairs. Afterward, bring the group back together to discuss the site. How did the use of video clips, sound, and photos enhance its content? What communication tools (e.g., still camera, video camera, or tape recorder) does the group want to use to document the soil experiments? Could the soil experiments be documented using graphs and charts, with drawings, or with interviews with someone who knows a lot about soil or underground life? Record the students' ideas on a whiteboard or a large sheet of paper.

Activity 5: Journal Activity

Suggested Exercises

- Imagine that you are a scientist going on a week-long trip to explore a cave deep underground. In words and drawings, show what you would bring for the trip.

- Imagine that you are an archeologist going on a week-long trip to uncover an ancient city buried in the desert. In words and drawings, show what you would bring for the trip.

Session 5
Field Trip

[Note: This session will take a minimum of two hours.]	
Preparation	Read through the Underground Network site (www.fieldmuseum.org/ua/nettop.htm) to ensure you understand all the activities for the field trip. • Make a contingency plan in case of unexpected events, such as bad weather. • Map out a schedule to ensure that you allot sufficient time for all activities.
Materials and Equipment	**For Testing Soil Structure** • Soil sample • Spray bottle of water • A quart-size jar with a lid or a 2-liter bottle • A funnel, if using a 2-liter bottle • A pencil or small stick • A ruler • An empty soup can with both ends removed • A watch with a second hand • Permanent markers • Clipboards, field journal worksheets, and extra paper • Pencils and pens • Optional: Pure samples of sand, silt, and clay for comparison (contact your local Soil and Water Conservation District for information on how to get samples) **For Measuring Scientific Variables** • A Celsius thermometer • A soil thermometer, if available (can be purchased at garden stores) • A cardboard tube (can use a paper towel roll) • Clipboards, field journal worksheets, and extra paper • Pencils and pens **For Drawing and Writing About Organisms** • A metric measuring tape • Five sticks, stakes, or pencils • A piece of string or yarn at least 2.5 meters (8 feet) long with a loop tied at one end • Two permanent markers of different colors • Trowels or spoons • Paper plates or white paper • Hand lenses • Margarine tubs or glass jars with holes poked in the top • Small Ziploc bags • Clipboards, field journal worksheets, and extra paper • Pencils and pens

(continued on next page)

Materials and Equipment *(continued)*	**General** • A camera (preferably digital) that can take still photos • A camera that can record short video clips • A hand-held tape recorder (to record people talking as well as insect or animal sounds) • A copy of *One Small Square Backyard* or a similar book with a detailed guide to different kinds of insects, worms, and animals

Activity 1: Pre-Field Trip Activities

Make sure that the group has a clear focus and clearly stated goals on the day of the trip.

• Discuss safety and appropriate conduct.

• Assign partners or have students choose them.

• To ensure that all the activities are covered, make a list of the activities and either assign or have partners volunteer to do specific ones.

• Assign special tasks, such as taking photos and recording video clips.

• Discuss how the equipment and materials will be shared.

• Review the field trip journal worksheets and the activity descriptions as a group.

Activity 2: Field Trip

Follow the directions for the field trip as described on the Underground Network site (www.fieldmuseum.org/ua/nettop.htm).

Session 6
Organize Data

Goals for the Session	• Transfer the data collected from the field trip into digital formats.
Preparation	• Review information on how to use your software programs to make charts and graphs.
Materials and Equipment	• A book to read aloud • Data from the soil experiment activities • Microsoft Word or a similar program • Software to make charts and graphs (Microsoft Excel or a similar program) • Software to edit photos (Adobe Photoshop Elements, PhotoDeluxe, or a similar program) • Newsprint or flip-chart paper • Colored markers • Composition books • Pens

Activity 1: Community Builder

Choose from the community builders described for earlier sessions, use one listed on the YouthLearn site (www.youthlearn.org), or develop your own.

Activity 2: Read-Aloud

Read aloud from any of the books listed on page 127, or add one of your own.

Activity 3: Review, Refine, and Digitize Data

Use software to turn the data collected from the soil experiment activities into charts, graphs, diagrams, and edited graphics.

1. Before starting, have group members look over their data to determine whether everything they wanted to accomplish with the experiments is complete. If someone lost some information, forgot to take an important photo, took a lot of bad photos, did not get to complete an experiment, or needs to redo an experiment, you may first want to devote attention to collecting the missing information.

2. The data collected from the experiments should include numerical information (such as temperatures and measurements) and graphical information (such as drawings, photos, and writing). Organize the information as follows:

   • Writing should be done on a computer using a word processing program and edited for spelling and grammar.

   • Diagrams about the organisms collected can be created with Inspiration or a similar program.

   • Digital photos can be edited with software such as Adobe Photoshop or PhotoDeluxe.

   • If a scanner is available, photos taken with a Polaroid or regular camera can be scanned. Drawings can also be scanned.

3. Have kids work in pairs with their own data. If access to computers and/or software is limited, create charts and photo and drawing displays on paper instead.

Activity 4: Journal Activity

Suggested Exercises

• In words and pictures, show how you would save soil and underground life from being harmed by human farming, mining, and construction.

• In words and pictures, show your model for a device that would recycle and save soil.

9

Session 7
Multimedia Presentation

Goals for the Session	• Learn how to storyboard. • Learn the tasks and roles involved in creating a multimedia presentation.

Preparation	Select software for the group to use to make the presentation. Your choice should be determined by what is available, what you feel comfortable using, and what you think the group is ready to use.
	
	    • If the group has more 9-year-olds than 11-year-olds, the slide show application in KidPix might be a good choice; it is easy to use and the kids will probably be familiar with it. (Children older than age 10 or 11 usually consider KidPix too childish.)
	
	    • Although not as simple as KidPix, HyperStudio is easy to learn, has more tools and functions than KidPix, and provides a good foundation for learning to use PowerPoint. If HyperStudio is available, it is probably the best choice for the 9- to 11-year-old age group.
	
	    • PowerPoint is a Microsoft Office application that usually comes with computers that have Microsoft Word and Excel. One of the major differences between PowerPoint and HyperStudio is that HyperStudio includes drawing tools that can be used to create original drawings and graphics right in the program. Otherwise, PowerPoint and HyperStudio are comparable in terms of tools and functions. HyperStudio is much more user-friendly and is therefore used by many elementary and middle school teachers. Using PowerPoint, however, has the advantage of giving the group experience with professional software used in the working world.
	
	Decide whether you want the kids to make one presentation as a group or multiple presentations in groups of two to four. Those who work well independently and who have created much data and many documents over the course of the project will probably benefit from working in smaller teams on separate presentations. Kids who work better with direct supervision will benefit from working on presentations as part of a group and with close adult supervision.

Materials and Equipment	• A book to read aloud • All the documents and images created with the soil experiment data and any other data or documents developed during the previous sessions (writing, drawings, photos, charts, or graphs) • Software for making a multimedia presentation (KidPix, HyperStudio, or PowerPoint) • Adobe Photoshop, Adobe PhotoDeluxe, or a similar application • Microsoft Word or similar application • Newsprint sheets, flip-chart paper, or a roll of butcher paper • Loose-leaf paper • Pencils and pens • Colored markers

Activity 1: Community Builder

Choose from the community builders described in earlier sessions, use one listed on the YouthLearn site (www.youthlearn.org), or develop your own.

Activity 2: Read-Aloud

Read aloud from any of the books listed on page 127, or add one of your own.

Activity 3: Storyboarding

Storyboarding is a technique for planning creative projects—such as movies and animation—that involve both narrative and visual elements. Multimedia presentations therefore lend themselves to storyboarding. In this exercise, students will storyboard their presentations either in teams (if they are making separate presentations) or as a group (if they are making a single presentation).

Lay out sheets of newsprint, flip-chart paper, or butcher paper, with one page to represent each slide (if using KidPix or PowerStudio) or card (if using HyperStudio). Brainstorm the content for each page. Put very rough sketches on the pages, enough to give an idea of the final image. Put a few words on each page (such as "woman opening door"), enough to outline the writing. Have two or three kids work together to make final decisions about the order and layout of the pages.

Activity 4: Production

Give an overview of the steps involved in making the multimedia presentation. Ask for volunteers for roles or assign roles (e.g., one person in the group could be in charge of creating the list of websites used for reference, another could be in charge of pulling together all the graphics, photos, and drawings).

List the steps for working on the presentation. If the group is working on separate presentations, have team members decide among themselves who will work on each step. The steps are as follows:

1. Create a title card/slide for the presentation.

   • List the names of all the people working on the presentation.

   • Give the date.

2. Design a minimum number of cards for the presentation.

   • Put images on each card.

   • Put a written description on each card.

   • Include charts, graphs, and tables, if you have them.

   • Include site drawings, graphics, and photos (each with the name of the person who made it).

3. Create a bibliography card/slide.

   • List websites used to gather information.

   • List books used to gather information.

4. Edit your parts of the presentation using the checklist (see Activity 5).

Activity 5: Presentation Quality Checklist

Create a list of quality attributes to use to evaluate the presentation. Use the list to improve the style, accuracy, and organization of the presentation.

Suggested Resource

Go to the South Central Regional Technology in Education Consortium website for project-based learning checklists for grades 2 through 4 (4teachers.org/projectbased/24mlt.shtml) and for grades 5 through 8 (4teachers.org/project based/58mlt.shtml). These lists can be customized. You can also check the site for checklists about writing, science, and oral presentation projects.

Session 8
New Projects

Goals for the Session	To be determined by the group.
Step 1	**Multimedia Presentation** Continue working on the multimedia presentation in this session and in future sessions, if necessary. Have each individual or small group present the completed project in front of the whole group, and allow time for discussion of each presentation.
Step 2	**New Products** Start planning new products that can be made with the content used for the multimedia presentation. Continue to use community builders, read-alouds, and journal activities, as well as mapping and webbing activities, as you develop new sessions.
Sample Projects	• **Website.** Make a website about soil. (See Session 6 of the *Civil Rights Through a Lens* project on page 158 for suggestions on how to do this as a group.) • **Group journal.** Make a book that can be copied and distributed to other children, parents, and adults. Storyboard the book pages. • **Videos.** Make a short video about soil. For first-time video projects, the simpler the equipment, the better. Some digital cameras can record short video clips with sound (up to 60 seconds). Or you can use a standard, low-end 8mm digital or analog video camera to make longer videos. See YouthLearn's article on making videos for a description of how to make a 30- to 60-second "problem-solution" video using a digital camera, a clip lamp, and storyboarding. • **Green Maps.** Take a look at the Green Maps website (www.greenmaps.org) for suggestions, templates, and examples of how to make customized maps that display the location or distribution of resources within a neighborhood or across a community. For example, how many community gardens, parks, and playgrounds are in your students' neighborhood? Where can people buy fresh produce? Where are the waste and hazardous dump sites in the community? Maps can be made by hand or with software.

Notes

★ Civil Rights Through a Lens (for ages 12 to 14)

Goals of the Project

Civil Rights Through a Lens introduces 12- to 14-year-olds to tools and techniques that can be used to create original digital content about social issues that are important to them. This project is targeted to this age group because these social issues—and related current events, politics, and history—are common themes in these students' school-work, in the literature and media they are exposed to, and in their personal lives.

The main goal of the project is for young people to learn how to collect, analyze, and present factual and expressive information about social issues that matter to them. This could be in the form of a multimedia presentation, as outlined in this project, or it could be in the form of short (30-second to 60-second) videos, a community map project, or a "virtual slide show" using drawings on newsprint or posterboard. The students will examine news, arts, and biographical materials as a vehicle for understanding the similarities and differences between the personal and social changes they are facing and the challenges that youth in similar cultures and environments have experienced in the past. Through their work on this project, young people will learn to perform in a collaborative work environment, to present information in multimedia formats, and to use multimedia authoring software and the Internet.

Components of the Project

- **Reading**—you will read to the group and have group members read on their own during every session.

- **Writing**—the group will practice writing during every session, producing original journal entries as well as text content for multimedia products, interviews, and news articles.

- **Oral presentation**—each group member will practice speaking in front of the group, listening to others speak, and providing critical feedback.

- **Visual communication**—each member of the group will create and edit graphical images with software tools.

- **Multimedia/Internet production skills**—the group will learn how to develop original content in digital formats.

Questions to Investigate

Helping the group identify questions they want to investigate is the critical first step of any inquiry-based project. To guide this process, it is helpful for you to pose your own questions prior to starting the project.

Your questions could be broad, such as:

- What was the Civil Rights Movement?

- What were the circumstances that led up to it?

- Who were the key players?

- Was the Civil Rights Movement successful?

- How is America different today because of it? How might it be different today if it had not occurred?

or questions could be more specific to a particular topic, such as:

- What role did young people play in the Civil Rights Movement?

- What actions did people fighting for civil rights take to accomplish their goals?

- What did African-American students experience when schools were integrated?

- What are people doing today to continue the struggle for civil rights?

(continued on next page)

Books for the Project

In addition to the materials listed in the overview of this chapter, we recommend the following book for personal and group reading:

- *Witnesses to Freedom: Young People Who Fought for Civil Rights*, by Belinda Rochelle (Econo-Clad Books, 1999). Features the personal stories of young African Americans—children, teens, and college students—who were active participants in the Civil Rights Movement in the 1950s and 60s. [**Note:** This excellent book can be difficult to find. Try Amazon (www.amazon.com), Barnes and Noble (www.bn.com), or your local used book store, and give yourself at least a couple of weeks to get the book before starting this project.]

Books with similar themes can be added or substituted. We recommend that the reading materials feature characters and cultural settings that reflect the culture and ethnicity of the group doing the project. Consider adding the following:

- *Dear Mrs. Rosa Parks: A Dialogue with Today's Youth,* by Rosa Parks with Gregory J. Reed (Lee & Low Books, 1997). Features a brief biography of Rosa Parks, followed by letters she has received from young people and her replies. She deals gracefully with sticky questions, such as her opinions about Louis Farrakhan and O.J. Simpson.

- *Pride of Puerto Rico: The Life of Roberto Clemente,* by Paul Robert Walker (Harcourt Brace, 1991). A thoughtful biography of baseball legend Roberto Clemente. The author shows that Clemente worked hard for what he achieved, rather than simply relying on natural talent.

- *Eighth Grade Writers: Stories of Friendship, Passage, and Discovery by Eighth Grade Writers,* edited by Christine Lord (Merlyn's Pen, Inc., 1996). A great model to encourage young people to write creative and personal stories.

- *Hispanic, Female and Young: An Anthology,* by Phyllis Tashlik (Arte Publico Press, 1995). This book is the result of a school project undertaken by eighth-grade girls at a school in Spanish Harlem and their teacher, Phyllis Tashlik, who was disappointed by the lack of literary resources available for Latina teens. Together they read stories and poetry by Latina authors and wrote their own.

- Various books by Walter Dean Myers. Myers is an award-winning author who specializes in fictional literature about the lives of young African Americans. Most of his stories center on teenage male characters but would engage teenage girls as well.

- Various books by Virginia Hamilton. Hamilton is an award winner who has written poems and novels about young African Americans living in the past and present, which provide opportunities to discuss important but difficult topics such as the legacy of slavery in the United States.

Session 1
Intro to Photojournalism

Goals for the Session	• Reflect on the experiences of young people involved with civil rights activism. • Analyze journalistic photographs. • Learn how to use a digital camera. • Learn basic digital photo-editing techniques.
Preparation	• Thoroughly preview the Photojournalist's Coffeehouse (www.intac.com/~jdeck/index3.html) site. Some of the content deals with teen pregnancy and other sensitive topics. Viewing it in advance will prepare you to handle questions and comments that may arise and allow you to block any content that may be inappropriate. • If possible, prepare a document beforehand (either in Word or in a web publishing program like Netscape Composer, a component of the free browser Netscape Navigator) with hyperlinks to the sites to be reviewed. • Register with the New York Times online site in order to access the articles for use in Activity 6 (www.nytimes.com). (Registration is free.)
Materials and Equipment	• Internet access—World Wide Web and email • Digital camera(s) • Software for editing photos—Adobe Photoshop or a similar application • A copy of the book *Witnesses to Freedom: Young People Who Fought for Civil Rights,* by Belinda Rochelle (if possible, have enough copies for the whole group) • Newsprint sheets or a roll of butcher paper • Pens • Colored markers • Composition books—one for each group member and instructor

Activity 1: Community Builder
Sample Community Builder: Commonality Warm-up

Divide into groups of three or four. Each small group of students is to find the things they have in common. After five minutes each small group reports back to the whole group.

Activity 2: Chapter Book Read-Aloud
The goal of chapter reading is to engage group members in an extended reading experience that will spark their interest in reading on their own. The reading also provides an opportunity to introduce themes, information, and ideas that relate to learning activities to be conducted later. A successful chapter book has a story and char-

acters that are easy to follow as well as colorful language that makes reading aloud interesting.

Suggested reading is the first chapter of *Witnesses to Freedom: Young People Who Fought for Civil Rights*, by Belinda Rochelle. Read the chapter aloud. (If you have extra copies, pass them out so members of the group can read along; if not, pass the book around as you go so group members can take turns reading.)

Start a discussion about questions related to the reading. Have the kids break into pairs to discuss the questions and then report back to everyone. Or write the questions on a whiteboard or piece of paper, ask group members to write responses in their journals, and then discuss their answers.

Suggested Questions

- Many students still go to schools where most of the students are the same race and ethnicity. How does this compare to the situation of segregation Barbara Johns and the other African-American students faced at R. R. Morton High School in Virginia?

- Is a boycott an effective way to help change a negative situation? What other things do you think the students at R. R. Morton High School could have done to help change the conditions at their school?

Activity 3: Mapping

Mapping is a technique to generate, share, and display ideas. In this exercise, kids create hand-drawn maps made up of concentric circles. Drawings, single words, and phrases are used to represent and develop ideas.

The purpose of the first map is to get the group thinking about how social problems are portrayed in the media and to introduce the mapping technique, which will be used again in later sessions.

1. Lay out a sheet of newsprint or a large sheet of butcher paper on a table. Write the words "social issues" in the center of the sheet of paper, and draw a circle around it.

2. Ask "What social issues have we recently seen covered in the media?" Have kids call out single words and short phrases in response to the question. (Drawings can also be used.) Write the words around the outside of the circle. Have one person add a word or drawing to model for the rest of the group. Draw a circle that encompasses both the students' words and the central word, creating two concentric circles.

3. Ask another question, such as "What social issues are not covered in the media or are covered only a little?" This time, have the entire group add words to the map. Draw a circle around these new answers, creating a third circle.

4. For a fourth circle, ask "What tools could we use to create information about social issues we think should be covered?"

5. After demonstrating how to make the first map, break the class into groups of four to six to work on separate maps for other questions, like the following:

   Write the phrase *social issues* inside the first circle.

   - Question for the second circle: What are the biggest social issues facing our community right now?

   - Question for the third circle: How do we get information about these social issues?

   - Question for the fourth circle: How can we communicate information about these social issues to others?

If the group is small (fewer than 10 members), most maps can be done as one group.

Activity 4: Web Exploration

As a model for developing their own photo essays, have group members view websites with photo essays that address social issues. A suggested site is Photojournalist's Coffeehouse (www.intac.com/~jdeck/index3.html); the site contains a number of web-based photo essays and has links to other photojournalism sites.

Have the kids view the photojournalism sites in pairs or as a group. Ask them to take notes on particularly interesting pages. Then bring all the group members together to discuss their impressions. Ask them:

- Did the essays confirm opinions you already had?

- Did they make you think differently about an issue?

- Was anything surprising? Disturbing? Inspirational?

Activity 5: Digital Camera—Up, Down, Close, and Far Photos

This exercise introduces the proper care, handling, and operation of digital cameras along with basic photography techniques. If a digital camera is not available, a Polaroid camera can be used. Cameras that require film processing before the pictures can be seen are not recommended for this activity.

9

Before introducing photography activities, visit websites with information on the basic terms and techniques of photography, such as the Kodak Guide to Better Pictures (www.kodak. com/global/en/consumer/pictureTaking/index.s html). Alternatively, bring in pictures of your own, cut them out from magazines, or ask the group to bring them in.

1. Give the group a quick overview of the parts of the camera (e.g., lens, shutter button, view window) and their functions. Show students how to hold the camera, and emphasize use of the camera strap to prevent accidental breakage.

2. Show the group how to take a picture from different angles and distances. Simple terms such as "up," "down," "close," and "far" can be used to describe angles and distances.

3. Let the students practice taking pictures. Give them a specific assignment, such as having each person take two pictures in the room close and from an up angle and two pictures from a down angle.

4. Load the photos onto computers, and have each person pick one or two photos to show to the group. Ask the kids to explain why they took their photos the way they did and what impact their choice made.

Activity 6: Journal Activity

Have the group read "The Impact of Violent Images: Assessing the Role of Photojournalism in Relaying the News" (www.nytimes.com/ learning/teachers/lessons/981026monday.html), an article on photojournalism from the New York Times Learning Network. Ask the kids to write their thoughts about the article in their journals (individually or as a group). As the New York Times site instructs, ask them to "explore the role of photojournalism in relaying news stories that involve death and destruction." Then ask them to discuss the idea of printing graphic photographs related to news events, and have them evaluate photographs and their accompanying stories to determine their purpose and relevance.

Activity 7: Session Wrap-up

It is a good idea to wrap up each session with a discussion that focuses on what that session was about. The goal is to develop a habit of reflecting on the day's learning and accomplishments.

Sample Discussion Questions

- Name three things you will take away from today's activities. What do you want to learn more about? What was the most difficult or challenging?

- Imagine that you have to teach what we learned today to someone else. What would you change about the things we did? What would stay the same?

Optional Extension Activity

Continue experimenting with photography. Give the group a new photo assignment each session to experience different techniques involving light, composition, angles, distances, and effects. Work with different types of cameras. Webmonkey's article on how to make "cheap and cool photos" (hotwired.lycos.com/ webmonkey/99/23/index3a.html) describes techniques such as Polaroid transfers (hotwired.lycos.com/webmonkey/99/23/index3 a_page8.html?tw=design), which can make really interesting photos. If a scanner is available, nondigital photos can be scanned for inclusion in the Internet photo essays.

Session 2
Editing Photos for Effect

Goals for the Session	• Learn how to edit a photo digitally. • Evaluate photo essay websites.
Preparation	• Practice the basics of Adobe Photoshop or other photo-editing application so you feel comfortable teaching it. • If possible, prepare a document beforehand (either in Word or in a web publishing program like Netscape Composer, a component of the free browser Netscape Navigator) with hyperlinks to the sites to be reviewed in Activity 4. • Choose and read an article to discuss in Activity 5.
Materials and Equipment	• Internet access—World Wide Web and email • A copy of the book *Witnesses to Freedom: Young People Who Fought for Civil Rights,* by Belinda Rochelle • Website evaluation sheets (see Activity 4) • Newsprint sheets or a roll of butcher paper • Pens • Colored markers • Composition books

Activity 1: Community Builder

Sample Community Builder: Hot Potato

First divide the class into groups of four or more. Then give each group a potato or other small object and start playing music. The potato is passed clockwise around the group until the music stops, at which point the person left holding the potato stands up and is asked to do an action—for example, clucking like a chicken. Once the person has done this, he or she can sit back down. The music starts again, and the kids resume passing the potato. The catch is that each person who has been given an action must stand up and perform that action every time he or she gets hold of the potato, even if only to pass it along to the next person. The result is a group of people doing different actions as the potato is passed.

Activity 2: Chapter Book Read-Aloud

Suggested reading is the second chapter of *Witnesses to Freedom.* Instructors and students read aloud as described for Session 1.

Suggested Questions

• Why do you think African-American families turned to the Supreme Court as the place to fight for changes in the public schools? What else could have been done?

• Spottswood says that as a result of the pressure on his family and the other families in the lawsuit to desegregate the schools, "I was expected to be perfect, and there's no such thing as a perfect child." How do you think you would have handled that situation if it had been you?

Activity 3: Photo Editing

Introduce software that can be used to alter photographs. Adobe Photoshop is the most powerful image-editing tool, but it is expensive and can be overwhelming. There are many other image-editing programs that are still powerful and yet cheaper and easier to use, including Adobe's new Photoshop Elements software. Whichever program you choose, demonstrate how to use its tools a few at a time.

9

1. Working with one of the students' photos from the previous session, demonstrate how to use several editing tools in Adobe Photoshop or Photoshop Elements. Try the following tools: replicating an image (the stamp), cutting an image (the crop tool), rotating an image, and zooming in on part of an image (the magnifier tool). Be sure to show how to save the original photo and work from a renamed duplicate. After demonstrating, have students work in pairs or individually to change the photos they have taken.

2. Bring the group back together to see an example of a specific photo alteration, and ask the group to try to reproduce it. The Mona Lisa illusion (www.exploratorium.edu/exhibits/mona/mona.html) from the San Francisco Exploratorium offers an example of several simple changes to the appearance of the Mona Lisa. Identify the changes and then try to make similar changes to one of your photos. Or download a Mona Lisa image and try to reproduce the Exploratorium's illusion. Check the Louvre (www.louvre.fr/louvrea.htm) website for Mona Lisa images to download.

Activity 4: Web Exploration

Have the group look at additional photo essay sites. Group members can work in pairs or individually. This time, ask everyone to pay attention to the design and navigation of the sites as well as the content. Have them use a website evaluation chart as a guide. The following sites have charts that can be adapted:

- Discovery Channel School, website evaluation chart for middle school students (school.discovery.com/schrockguide/evalmidd.html)

- Midlink Magazine, a web page evaluation form by Ligon Middle School (www.ncsu.edu/midlink/tutorial/WWW.eval.html)

We suggest the following photo essay sites:

- Seattle Times Civil Rights Photo Tour (seattletimes.nwsource.com/mlk/movement/PT/phototour.html). Features famous civil rights photos. Also links to a civil rights timeline and current news articles on civil rights-related topics.

- In Transit: Interviews and Photographs of People on the Bus (www.geocities.com/~in transit1/home/home_frm.html). Artist-created site consisting of interviews and photos of people using public transportation in the San Francisco Bay Area.

- YO! Youth Outlook Photography Workshop (www.pacificnews.org/yo/photo/index.html). Part of a news organization that produces and distributes articles, photography, and radio programming by teens.

Activity 5: Discussion

Read and discuss a news article written by or about teens reporting or taking action on issues in their community. YO! Youth Outlook (www.pacificnews.org/yo/index.html) has free articles that can be downloaded. In addition, LeAlan Jones and Lloyd Newman, two African-American teenagers in Chicago, created an award-winning documentary for National Public Radio called *Ghetto Life 101*. They created a second documentary, *Remorse: The 14 Stories of Eric Morse*, about two young boys who dropped a five-year-old child out of a 14th-floor window at the Ida B. Wells public housing development. Excerpts of the shows can be downloaded from the NPR website (www.npr.org) for free. Full tapes and transcripts can be ordered through the NPR website.

Activity 6: Session Wrap-up

Have the kids spend about 10 minutes writing in their journals about what they learned in this session, what questions it made them want to investigate, or how they feel about the session. See Session 1 for sample questions.

Optional Extension Activities

- Complete the Webmonkey Photoshop Crash Course (hotwired.lycos.com/webmonkey/98/20/index0a.html).

- Do another exercise with a new tool in Photoshop. The Webmonkey Photoshop Crash Course has easy-to-follow lesson plans on specific tools and techniques. Select one lesson to try. You might want to take a look at other sites with Photoshop lessons, such as the University of Washington Technical Communication Program site (www.uwtc.washington.edu/docdepot/Documents/photoshop/lessons/Default.htm) or the site for Ricks College in Rexburg, Idaho (www.ricks.edu/Ricks/employee/WILLIAMSL/whome.htm).

Session 3
Formulating the Inquiry and Starting the Project

Goals for the Session	• Define questions and issues to be investigated through the photo essay. • Practice interviewing skills. • Create inquiry sheets. • Create a visual map. • Create an options chart. • Create interview question cards.
Preparation	• Print out or prepare on newsprint the inquiry process handouts. • Consider using a KWHL chart as an alternative for defining the inquiry process (see page 129). • Prepare an outline of the inquiry chart for planning the site visit (see Activity 5).
Materials and Equipment	• Internet access—World Wide Web • A copy of the book *Witnesses to Freedom: Young People Who Fought for Civil Rights,* by Belinda Rochelle • Newsprint sheets or a roll of butcher paper • Index cards • Pens • Colored markers • Inquiry formation sheets (see Activity 3) • Composition books

Activity 1: Community Builder
Sample Community Builder: Fruit Basket

The group sits in a circle on chairs (with no extra chairs in the circle), except for one person who stands in the middle. The group members sitting in chairs are equally divided among three fruits (e.g., apples, oranges and pineapples). The middle person then calls out the name of one of the fruits (e.g., apples), and all the apples plus the person in the middle change seats (who then becomes an apple, or whichever fruit he or she called). The person left without a seat becomes the next caller. If a caller says "fruit basket" all players must change seats.

Activity 2: Chapter Book Read-Aloud

Suggested reading is the third chapter of *Witnesses to Freedom.* Instructors and students read aloud as described for Session 1.

Suggested Questions

• What is happening in the photo of Elizabeth Eckford walking to Central High School in Little Rock, Arkansas? How does this photo make you feel? Why do you think the photographer chose this particular composition (arrangement of elements in the photo)?

• Elizabeth Eckford and the other eight African-American students at Central High School experienced a lot of abuse every day. Why was it important for them to respond to the abuse in a nonviolent way? What do you think would have happened if they had responded with verbal or physical violence?

• Elizabeth Eckford says, "Even though there was a screaming mob outside of that school every day, there were a lot of people—families and people that I didn't know—who supported us." How did Elizabeth and the other eight African-American students know they had support?

Activity 3: Inquiry Sheets

In this exercise, the group will write questions to define the topics they want to investigate in their photo essays.

1. Print out the inquiry process handouts from the YouthLearn website, or make your own handouts. All group members should have the following sheets:

  • A sheet where they can record what they want to know and what they think they already know (www.youthlearn.org/learning/images/illus/1-1-1_1.gif). Headings can include:

    ■ I WANT TO KNOW…

    ■ I ALREADY KNOW…

    ■ I DON'T KNOW…

    ■ I THINK MAYBE…

  • A sheet where they can refine their questions (www.youthlearn.org/learning/images/illus/1-1-1_3.gif):

    ■ MY MAIN QUESTION…

    ■ ANOTHER QUESTION…

    ■ ANOTHER QUESTION…

    ■ ANOTHER QUESTION…

The idea is to continue to refine the questions until the ones that best lend themselves to investigation are uncovered.

2. Have kids work in pairs on their question sheets. Bring the group back together to

have everyone share questions. Write the questions on cards or a sheet of paper and post them.

Option for Defining the Inquiry Process

Another way to define the inquiry for the photo essay content is to create a KWHL chart. K stands for "What do we KNOW?" W stands for "What do we WANT to find out?" H stands for "HOW can we find out what we want to learn?" L stands for "What have we LEARNED?" (See Session 1 of the sample inquiry-based project *The Soil Around Us* for an illustration of a KWHL chart.)

Activity 4: Mapping—Preparation for Site Visits

Map ideas for a site visit to collect content for the photo essay. Taking photos would be one aspect of the visit; other activities might be interviewing and conducting surveys. A photo essay site could be as close as a space inside your building, the street outside your door, or a park in your neighborhood. Or the site could be farther away, requiring transportation and a scheduled visit. Scheduling interviews requires making appointments with interviewees in advance of the visit and ensuring that they are aware of the photo essay topic.

• Question for the second circle: What sites can we visit to take photos about the questions and issues we want to investigate?

• Question for the third circle: What other activities can we do at these sites to create content for the photo essays?

• Question for the fourth circle: Who at these sites can help us?

Activity 5: Options Chart

Have the group create and fill in a chart of the potential photo essay questions, content collection, and site locations to help make the final decision on what to do and where to do it.

Depending on the number of people, the group could choose to do one essay as a group or to do separate essays by working in pairs or teams of three or four. Working individually is not recommended because it will increase the time required to gather material and reduce opportunities for cooperative work. Time and resources permitting, the group may decide that multiple site visits are required to collect content for the essay.

Activity 6: Interview Role-Plays

Interviewing will be an integral part of collecting the photo essay content, and successful interviewing requires practice. In this exercise, group members brainstorm interview questions and role-play an interview.

1. Have the group members work in pairs. Each pair should use index cards to write down five or six interview questions about their photo essay topic.

2. Reconvene the group to role-play interviewing someone using the questions they have created. First, ask a volunteer to do a role-play with the instructor. The instructor should be the interviewer first, then the volunteer should be the interviewer. Point out what was effective about the volunteer's interviewing techniques and make suggestions for improvement. Emphasize body gestures, volume and clarity of speech, and effectiveness of the questions. Note-taking techniques are also important.

3. Ask two new volunteers to role-play an interview in front of the group.

4. Have the pairs practice on their own.

Tip

If the students would like to conduct a survey as part of content collection, ask them to brainstorm survey questions and have them choose the questions they like the most. When the questions have been chosen, create a role-play so the kids can practice conducting the survey. After the role-play, encourage the kids to refine the questions.

Activity 7: Personal Reading

Group members select books or other reading material to read silently for at least 10 minutes.

Activity 8: Session Wrap-up

Have group members spend about 10 minutes writing in their journals about what they learned in this session, what questions it made them want to investigate, or how they feel about the session. See Session 1 for sample questions.

Sample Inquiry and Site Visit Chart

Issue	Questions	Possible Content Sources	Possible Sites
• Race/Identity • Education	• Are our schools still segregated? • What experiences have you had from which you learned something important about racial issues, challenges, conflicts, etc.?	• Interview students • Survey students • Interview journalist covering education issues • Interview people who graduated from our school district many years ago • Photos of all interviewees • Photos of school or location • Photos that reflect the focus of our inquiry	• Middle school • Youth program • Homes

Session 4
Writing Skills

Goals for the Session	• Practice different modes of writing. • Create a checklist to assess writing.
Preparation	• Familiarize yourself with the lesson "Making the Personal Political: Writing Opinion Pieces About Meaningful Issues to Kids" (www.nytimes.com/learning/teachers/lessons/000414 friday.html) from the New York Times Learning Network for use with the group in Activity 3. • Print out a copy of the lesson and multiple copies of the related *New York Times* article "Growing in the Job" (www.nytimes.com/learning/general/featured_articles/000414friday.html). • Create a writing checklist using the resources suggested in Activity 4.
Materials and Equipment	• Internet access—World Wide Web • A copy of the book *Witnesses to Freedom: Young People Who Fought for Civil Rights,* by Belinda Rochelle • A copy of the New York Times Learning Network lesson plan "Making the Personal Political: Writing Opinion Pieces About Meaningful Issues to Kids" (www.nytimes.com/learning/teachers/lessons/000414friday.html) • Copies of the New York Times Learning Network news article "Growing in the Job" (www.nytimes.com/learning/general/featured_articles/000414friday.html) • Copies of descriptions of writing modes (see Activity 4) • Copies of the writing checklist (see Activity 5) • Microsoft Word or a similar application • Newsprint sheets or a roll of butcher paper • Index cards • Pens • Colored markers • Composition books

Activity 1: Community Builder

Sample Community Builder: Stand Up

Have your group get in pairs. The pairs should sit on the floor, back pressed to back, and try to stand up without using their hands. After a pair stands up, have them find another pair, and all four of them must sit down and stand up. Continue until the entire group is trying to stand up together. This game promotes friendliness and is especially good with a large group.

Activity 2: Chapter Book Read-Aloud

Suggested reading is the fourth chapter of *Witnesses to Freedom.* Instructors and students should read aloud as described in Session 1.

Suggested Questions

• Do you think that Claudette's refusal to give up her seat was an act of leadership? If not, what was it? Why did she do it?

• Claudette described herself as "feeling like an outsider." What made her feel this way? Do you think that feeling this way made what she did more or less courageous?

• Claudette's refusal to give up her seat became part of the Montgomery bus boycott organized by Martin Luther King, Jr. Why do you think a boycott was again chosen as the form of "nonviolent protest" to help change the laws? How would the effects of a bus boycott be different from boycotting a school, as Barbara Johns and the other teens did in Virginia?

Activity 3: Writing Opinion Pieces

Using the lesson "Making the Personal Political" from the New York Times Learning Network and the related article "Growing in the Job" (see Preparation section for web addresses), help each group member choose a controversial topic and write a persuasive opinion piece. Have them model their own pieces after "Growing in the Job."

Activity 4: Writing Modes

In this exercise, you discuss the different modes of writing with the class and do writing prompts to practice some of those modes. The goal is to be able to identify different writing modes in order to select a writing mode for the text of the photo essays.

1. Go to the Custer Elementary School (WA) website for a description of the five basic writing modes (www.ferndale.wednet.edu/custer/Writing/modes.html): descriptive, narrative, imaginative, expository, and persuasive. Post this list on a large sheet of paper, or make copies for the group. Discuss the different kinds of writing with the students, noting that the *New York Times* writing exercise they just completed is an example of persuasive writing.

2. Select writing prompts for at least two modes of writing other than persuasive, and have the group spend 10 minutes writing in each of the modes.

3. Have the group members share and discuss their writing with a partner. Invite partners to comment on the written pieces. Ask them to focus on what they consider to be the most engaging parts.

Activity 5: Writing Checklist

This exercise involves creating a checklist of guidelines for high-quality writing. The group will use the guidelines for future editing of written content.

1. Go to the South Central Regional Technology Consortium's Project-Based Learning Writing Checklist for Grades 5-8 (4teachers.org/projectbased/58wrt.shtml). Print out copies of the list for the group.

2. Review the list with the group, and ask them for suggestions on guidelines to add or remove.

3. When final decisions have been made, return to the checklist page and reprint the list with the changes.

Activity 6: Personal Reading

Kids select books or other reading material to read silently for at least 10 minutes.

Activity 7: Session Wrap-up

Have the group spend about 10 minutes writing in their journals about what they learned in this session, what questions it made them want to investigate, or how they feel about the session. See Session 1 for sample questions.

Session 5
Site Visit

Goals for the Session	• Take journalistic and artistically expressive photos. • Conduct and record interviews.
Preparation	The site visit(s) to collect content for the photo essays should be coordinated as carefully as a full-blown field trip. • Make a contingency plan for unexpected events, such as bad weather, that may disrupt outdoor photo taking. • If the visit will be to a site outside your program building, discuss safety and appropriate public conduct. • Map out a schedule to ensure that sufficient time is allotted for interviewing, taking photographs, and other activities. • IMPORTANT: If interviews are going to be conducted with specific individuals, instructors should talk to them beforehand about the goals of the photo essay and the intention to publish on the web for educational/nonprofit purposes. Get interviewees' permission to take photos and record interview content. If possible, have interviewees and people who have agreed to be photographed sign a release form. Midlink Magazine (www.ncsu.edu/ midlink/posting.html) has links to copyright guidelines and permission-request forms for posting and linking to published materials, such as graphics, photos, and text.
Materials and Equipment	• Adobe Photoshop or similar application • Microsoft Word or similar application • Digital camera(s) • Release forms and copyright permission forms, if applicable (see Preparation section above) • Interview question cards • Clipboards and paper (easier to carry than notebooks) • Pens and pencils • Index cards • A hand-held tape recorder for interviews (optional)

Site Visit to Capture and Create Focused Photo Essay Content

This session has a different structure than the others so far—you'll notice there are no community builders or read-aloud activities built in. This session focuses on using the skills the group members are learning to start collecting information for the group's photo essay.

Make sure that group members have a clear focus and goals on the day of the visit:

• Review the key questions of the photo essay inquiry.

• Discuss and assign roles to ensure clarity about who will assume responsibility for

conducting interviews, recording interviews, taking notes, taking photos, and other tasks.

• Discuss how the camera(s) and any other equipment will be shared.

• Decide on target goals for the trip, such as a minimum number of photos to be taken and a minimum number of interviews to be conducted.

• Carry out the site visit according to plan; take photos and conduct interviews, keeping a log of each.

• On return from the site visit, begin typing up interview notes and editing photos.

Session 6
Web Publishing, Part I

	[Note: The activities in this session can be spread across several sessions if time permits.]
Goals for the Session	• Learn how to storyboard. • Learn the roles and tasks involved in planning, designing, and building a website.
Preparation	If you have never published a website before, give yourself a sample project and put up a small website consisting of at least three or four linked pages. Try using one of the web development guides that the group will use, such as Lissa Explains It All or Webmonkey for Kids (see Part 3 of this session for the web addresses). or Prepare the storyboard as instructed, but plan to present the final product as an issue-focused bulletin board. Be sure to include a clear and strong visual introduction to the issue as well as photos and written commentary. Have the group find a strong quote for the conclusion.
Materials and Equipment	• Internet access—World Wide Web and email • A copy of the book *Witnesses to Freedom: Young People Who Fought for Civil Rights*, by Belinda Rochelle • Adobe Photoshop or similar application • A web publishing program—Netscape Composer, Dreamweaver, or a similar application • Other web publishing tools as needed—see suggestions in the web development guides Lissa Explains It All and Webmonkey for Kids • Microsoft Word or similar application • Newsprint sheets or a roll of butcher paper • Loose-leaf paper • Pens and pencils • Colored markers • Composition books

Activity 1: Community Builder

Choose from the community builders in earlier sessions, use one listed on the YouthLearn site (www.youthlearn.org), or develop your own.

Activity 2: Chapter Book Read-Aloud

Suggested reading is the fifth chapter of *Witnesses to Freedom*. Instructors and students read aloud as described for Session 1.

Suggested Questions

• What is happening in the photo of the lunch counter sit-in? What do the faces of the nonviolent protesters at the counter tell you about their feelings? What do the faces of the crowd around the counter express? Why do you think the photographer chose this particular composition (way of arranging the elements in the photo)?

- Why do you think the high school and college students chose places like the Woolworth lunch counter to stage sit-ins?

- Do you think the students were right to train for the sit-ins without telling their families? What were the advantages of doing this? What were the dangers?

Activity 3: Web Development Roles

Give an overview of the components of web development and publishing. Assign roles.

Before starting, do the following:

- Make sure that group members have finished typing up their notes from the site visit and have loaded and checked out the photos.

- Be sure to review information on copyright guidelines for web publishing. Midlink Magazine (www.ncsu.edu/midlink/posting. html) has links to copyright guidelines and permission-request forms.

- Select a web development guide (see below) that kids can use for instructions and advice while they plan and build the web pages for their photo essays.

Suggested Web Development Guides

- Lissa Explains It All (www.lissaexplains.com/basics.shtml). Thirteen-year-old Lissa started this guide when she was 11 to help other kids learn how to design and program websites. The site is easy to understand, is full of detail, and has lots of links to free or inexpensive software, Internet tools, and services.

- Webmonkey for Kids (hotwired.lycos.com/ webmonkey/kids/lessons/index.html). This is a kids' version of the excellent Webmonkey site for web developers. The Projects section has step-by-step instructions for building sample pages, and the Slide Show project (hotwired.lycos.com/ webmonkey/kids/projects/slideshow.html) is a good model for photo essays.

- Webmonkey also has a planning guide (hotwired.lycos.com/webmonkey/kids/plan ning/index.html) with tips for adult

instructors on how to introduce web development. Be sure to check out the Advice from a Teacher section, which has practical organizing suggestions (e.g., creating a desktop folder for each group member to store work and creating shortcuts for the software that will be used).

Activity 4: Develop Working Teams

1. List the roles involved with web publishing (see below), and post them where everyone can see them. Define and organize the roles in a way that makes sense to you and that you think will work for the group.

Suggested Roles for Working Teams

- Layout and Design: Decide how the visual elements of the web pages—graphics, text, links—will be arranged on each page. Choose the colors, font, and font size for each page. The layout can be storyboarded (see Activity 5).

- Architecture: Decide how the pages will be linked together. Use software to make a diagram of how the pages will be connected, or use paper and pencil. Inspiration K-12 is great for making diagrams. Microsoft Word has box, circle, and line tools that can be used to make diagrams.

- Graphics: Create original images or logos, etc. for the pages. Change or edit existing images as needed.

- Coding/Programming (if you have advanced students): Use web publishing software to assemble the web pages. Write HTML tags, tables, frames, scripts, and other code and program elements. Work with layout and design, architecture and graphics team to ensure that pages load quickly and easily and that site navigation works.

- Editing: Review all written content to check for errors. Check copyright guidelines to make sure that all text and image content is proper to use and appropriately attributed.

9

Additional Roles

- Identify and secure web publishing space (if needed).

- Identify and load web publishing tools (if needed).

- Design pages, such as the front page, that are not specific to any one person's content.

2. Discuss the roles with the group. Have kids select a role or multiple roles, weighing their interests and skills for the specified tasks.

3. In addition to taking on a role or roles in the development of the site, each group member will need to assume responsibility for providing a final version of the text and photo content he or she will be contributing to the photo essay.

Activity 5: Storyboarding

Storyboarding is a technique used to plan creative projects that involve both a narrative and visual elements, such as movies, animation, and picture books. Websites also can be storyboarded.

1. Lay out sheets of newsprint or butcher paper.

2. Brainstorm what will happen on each web page. Put rough sketches on the pages, enough to give an idea of what the final picture will be. Write a few words on each page (e.g., "woman opening door"), enough to outline what the writing will be.

3. After the pages are made, have two or three people work together to make final decisions about the order and layout of the pages.

Activity 6: Build Web Pages

Start building web pages using the web development guide you chose earlier in this session.

Activity 7: Personal Reading

Group members select books or other reading material to read silently for at least 10 minutes.

Activity 8: Session Wrap-up

Have group members spend about 10 minutes writing in their journals about what they learned in this session, what questions it made them want to investigate, or how they feel about the session. See Session 1 for sample questions.

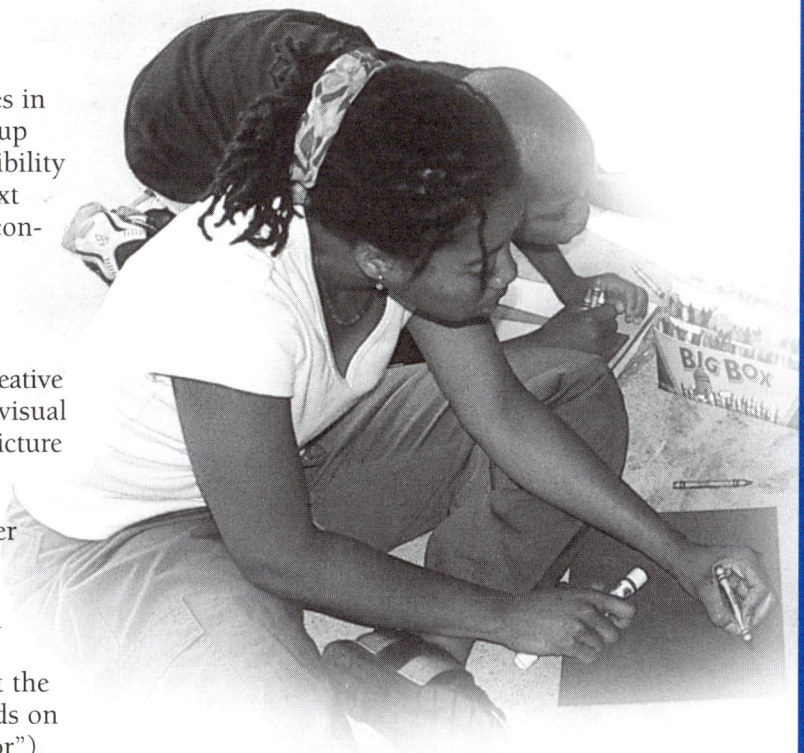

Session 7
Web Publishing, Part II

Goals for the Session	• Practice web development skills. • Learn effective linking of photos and writing. • Plan, implement, and critique the presentation of the project.
Preparation	• Customize the quality checklist to suit web-based photo essays (see Suggested Resources below).
Materials and Equipment	• Internet access—World Wide Web and email • A copy of the book *Witnesses to Freedom: Young People Who Fought for Civil Rights*, by Belinda Rochelle • Adobe Photoshop or similar application • A web publishing program—Netscape Composer, Dreamweaver, or a similar application • Other web publishing tools as needed—see suggestions in the web development guides Lissa Explains It All (www.lissaexplains.com/links.html) and Webmonkey for Kids (hotwired.lycos.com/webmonkey/kids/tools/index.html) • Microsoft Word or similar application • Newsprint sheets or a roll of butcher paper • Loose-leaf paper • Pens and pencils • Colored markers • Composition books

Activity 1: Community Builder
Choose from the community builders in earlier sessions, use one listed on the YouthLearn site (www.youthlearn.org), or develop your own.

Activity 2: Chapter Book Read-Aloud
Suggested reading is the sixth chapter of *Witnesses to Freedom*. Instructors and students should read aloud as described in Session 1.

Suggested Questions
• Why do you think whites and African Americans did the freedom rides together? How might this have made things easier—or harder—than if African Americans had done the rides on their own?

• What do you think Diane Nash means when she says, "Ending discrimination is not only a struggle to change laws. Internal liberation is just as important"?

• Can you think of a situation in your school, neighborhood, or community in which people from different groups have had to cooperate to achieve a common goal? What benefits came from the cooperation? What made the situation difficult?

Activity 3: Web Pages
Continue building web pages for the photo essay.

9

Activity 4: Website Quality Checklist

Create a list of attributes you would like to see the photo essay website incorporate. Use the list to improve the style, function, and navigation of the web pages.

Suggested Resources

- South Central Regional Technology in Education Consortium Project-Based Learning Checklist (4teachers.org/project based/58mlt.shtml). The checklist is for multimedia presentations but can be customized. Many of the items are relevant to website development.

- WWWCyberGuide Rating for Web Site Design (www.cyberbee.com/guide2.html). This easy-to-use form was developed by Karen McLachlan, a teacher at East Knox High School in Ohio.

- Yale University Web Style Manual (info.med.yale.edu/caim/manual). This comprehensive style guide is used by professional web developers.

Activity 5: Personal Reading

Group members select books or other reading material to read silently for at least 10 minutes.

Activity 6: Session Wrap-up

Have group members spend about 10 minutes writing in their journals about what they learned in this session, what questions it made them want to investigate, or how they feel about the session. See Session 1 for sample questions.

or

If appropriate, present completed web-based photo essay to another group (for example, parents, staff, or community members). Ask for evaluative feedback.

Session 8 and Beyond
Related Projects

Goals for the Session	• Continue working on the web pages during this session and future sessions, if necessary. • Start planning related projects.
Related Projects	• **Videos.** Revisit the inquiry and content created for the photo essay to make a short video with whatever digital recording equipment is available. For first-time film projects, the simpler the equipment, the better. Sony Digital Mavica cameras can record short video clips with sound, up to 60 seconds long. Or you can use a standard, low-end, 8mm digital or analog video camera to make longer films. See the section on Youthlearn about making videos for more information and for examples of video projects made by kids. • **Multimedia presentations.** Use the photo essay web page content to make a multimedia presentation with HyperStudio, PowerPoint, or a similar application. Begin by storyboarding each slide of the presentation. Have group members work in pairs to complete the slides. Record narration and other sounds on the slides. If the application can support imported HyperStudio files, think of ways to incorporate short video clips or animations into the presentation. Have two or three kids work together to make final decisions about the order and layout of the presentation. • **Green Maps.** Check the Green Maps website (www.greenmaps.org) for suggestions, templates, and examples of how to make customized maps that display the location or distribution of resources within a neighborhood or across a community. For example, where are health clinics and other health service providers located in relation to the residential location and density of youth in the community? What is the location of waste and hazardous dump sites in relation to the location of recreation areas for youth and families? Maps can be made by hand or with software.

Notes

Endnotes

1. The Pew Partnership for Civic Change, *Solutions for America: What We Know Works* (Richmond, VA: University of Richmond, 2001), p. 25 (www.pew-partnership.org/pdf/what_works/what_works_9-36.pdf).

2. The National Institute on Out-of-School Time, *Fact Sheet on School-Age Children's Out-of-School Time* (Wellesley, MA: Wellesley College, 2001), p. 3 (www.niost.org/fact_sheet_01.pdf).

3. The Pew Partnership for Civic Change, *Solutions for America*, p. 25.

4. The Pew Partnership for Civic Change, *Solutions for America*, p. 25.

5. Adapted from the wNet School workshop *After School Programs: From Vision to Reality* (www.thirteen.org/wnetschool/concept2class/month11/implementation.html).

6. Adapted from the wNet School workshop *After School Programs: From Vision to Reality* (www.thirteen.org/wnetschool/concept2class/month11/implementation.html).

7. Outcome Measurement section adapted from United Way of America, *Measuring Program Outcomes: A Practical Approach* (United Way of America, 1996), pp. 4-5 (national.unitedway.org/outcomes/publctns.htm#lt0989).

8. *Who Should Do the Evaluation?* and *Emphasize the Ongoing Nature of Evaluation* sections adapted from K. E. Walter, et al, *Beyond the Bell: A Toolkit for Creating Effective After-School Programs, Second Edition* (Naperville, IL: North Central Regional Educational Laboratory, 2001), pp. 13-14 (www.ncrel.org/after/bellkit.pdf).

9. Written for this guide by George Gundrey, CompuMentor, San Francisco, CA (www.compumentor.org).

10. Adapted from material written for the YouthLearn website (www.youthlearn.org) by Alnissa Algood, Nonprofit Tech Association, San Francisco, CA (www.nonprofit-tech.org).

11. Adapted from material written for the YouthLearn website by Alnissa Algood, Nonprofit Tech Association.

12. The Nonprofit Tech Association is just one of a quickly growing group of nonprofits whose purpose includes providing technical services to the nonprofit sector. Nonprofit Tech currently has offices in the San Francisco Bay area and Ann Arbor, MI. Other agencies that offer an array of technical services for nonprofits include CompuMentor (San Francisco, CA); NPower (Seattle, WA); NetCorps (Eugene, OR); Technology Works for Good (Washington, DC); and the Center for Management Assistance (Kansas City, MO).

13. Adapted from material written for the YouthLearn website by Jayne Cravens, United Nations Volunteers, Bonn, Germany (www.unv.org).

14. Several exceptions to the rules exist; for more information about the FTC's rules, visit the FTC's website (www.ftc.gov/bcp/conline/edcams/kidzprivacy/index.html).